MECHANICS

This book has been written designedly for the private student and especially for those who need to acquire a knowledge of the mechanical principles which underlie much of the work in many branches of industry. SI units are used throughout.

TEACH YOURSELF BOOKS

MECHANICS

P. Abbott, B.A.

Revised by
D. G. Kershaw, M.A.

TEACH YOURSELF BOOKS
Hodder and Stoughton

First printed 1941
Revised edition 1971
Fifth impression 1976

ISBN 0 340 05653 3

Printed in Great Britain
for Teach Yourself Books, Hodder and Stoughton, London
by Richard Clay (The Chaucer Press), Ltd., Bungay, Suffolk

INTRODUCTION

This book has been written designedly for the private student and especially for those who need to acquire a knowledge of the mechanical principles which underlie much of the work in many branches of industry and in the technical branches of the fighting forces. It is difficult in these troubled times to obtain the tuition provided by Scientific and Technical Institutions for such subjects as Mechanics. Nor is there any completely satisfactory substitute. Nevertheless the author hopes and believes that enough can be learnt from the present volume to be of very real practical use as an introduction to the subject. Like other volumes in the "Teach Yourself" series, it seeks to give such help as is possible to those students who are anxious to acquire a knowledge of the subject but must rely, in the main, on their own studies.

The difficulty is increased in such a subject when access is not possible to laboratories and apparatus. The view of the author of this book, founded on long experience, is that the approach to mechanics should be largely experimental. To a considerable extent the subject has been built up through centuries of human progress by experiment, often prompted by the needs of mankind. Experiment has preceded theory, as witness the discoveries of Archimedes, Torricelli and Galileo.

What is to take the place of this practical basis for the subject? In the early chapters of this book directions and descriptions are given of experiments with simple apparatus such as most students with a little ingenuity can construct or obtain. By this means the student is led to formulate some of the simple fundamental principles upon which the subject is built.

When the apparatus required is more complicated and cannot be obtained by the student it seems to the author that in many cases there is little value in the description of experiments which have been performed by others, with the numerical results which must be accepted without personal verification. It seems simpler and just as convincing to state the principles which are demonstrated by such experiments, leaving the verification for later study.

Mathematical proofs which involve a knowledge and experience of the subject which may be beyond the average student are omitted, the truth of the principles which they demonstrate being assumed.

Another important factor in a book of this kind is that of the amount of mathematical knowledge which may be assumed as being possessed by the average reader. The minimum amount required for the greater part of this volume is ordinary arithmetic, elementary algebra, including the solution of quadratic equations, with a knowledge of ratio, variation, and a certain amount of fundamental geometry. In some sections of the book a knowledge of elementary trigonometry is essential. To assist the student cross references, when it seems desirable, are given to the appropriate sections in the companion book on Trigonometry in this series. In those cases in which practical drawing can be employed as an alternative to a trigonometrical solution, this is indicated.

Inevitable limitations of space have made it necessary to exclude many practical and technical applications of the principles evolved. It has been considered more profitable for private students, who in many cases are familiar with practical applications, that as much space as possible should be given to explanations of the theoretical aspects of the subject.

The examples to be worked by the student are designed to enable him to test his knowledge of the theorems on which they are based and to consolidate his knowledge of them. Academic exercises, depending for their solution mainly on mathematical ingenuity, have been excluded.

The author desires to express his thanks to Mr. W. D. Hills, B.Sc., for permission to use diagrams 22 and 138 from his book on *Mechanics and Applied Mathematics*, published by the University of London Press; also to Messrs. Cussons for the use of blocks of some of the admirable apparatus which has been designed by them for use in teaching mechanics. The author is also indebted to Mr. C. E. Kerridge, B.Sc., for the use of the example on p. 237 from *National Certificate Mathematics*, Vol. I, and to the University of London for their consent to the inclusion of a few of their Examination questions.

1941

PUBLISHER'S NOTE TO 1971 EDITION

In order to bring it completely up to date and to incorporate SI units, this edition has been extensively revised, and in some parts re-written, by David Kershaw. Harold Frayman has metricated the exercises.

CONTENTS

CHAPTER I

Introductory

CHAPTER II

The Lever

CHAPTER III

Centre of Gravity

CHAPTER VIII

Bodies in Motion; Velocity

CHAPTER IX

Acceleration

CHAPTER X

Newton's Laws of Motion

CHAPTER XI

Bodies in Collision

CHAPTER XII

Gravitation, Mass and Circular Motion

CHAPTER XIII

Work, Energy and Power

CHAPTER XIV

Machines

CHAPTER XV

Composition of Velocities; Relative Velocity

CHAPTER XVI

Projectiles

CHAPTER XVII

Density and Relative Density

CHAPTER XVIII

Liquid Pressure

CHAPTER XIX

The Pressure of Gases

CHAPTER I

INTRODUCTORY

1. The meaning of mechanics

The word "Mechanics" is derived from a Greek word meaning "*contrivances*", and this conveys some idea of the scope of the subject. Many of the "contrivances" and the fundamental principles underlying them form part of the instinctive heritage of the human race. Our body contains some of these contrivances, which, through long ages, have been adapted by Nature to our needs. If we lift a weight, or raise our feet from the ground in walking, we are employing "mechanisms" which are admirably adapted by nature for the purpose. A little child, in learning to stand and walk, is contriving by his experiments in balancing to solve problems which later will come up for our consideration in this book. Two children on a "see-saw" know the necessity of positioning themselves so that they may balance.

The principles of mechanics also enter vitally into the daily work of many, whether it is the bricklayer wheeling his barrow-load of bricks, the farmer pumping water from his well, or the aeronautical engineer designing an aeroplane. The scientific study of these principles through the ages has led to the complicated mechanisms of modern times, such as the steam-engine, the motor-car, or the machine-gun. It is important, therefore, that we should examine and learn to understand them, and that is the principal reason for studying mechanics.

2. Mechanics may be defined as the subject in which we study *the conditions under which objects around us move or are at rest.*

These two aspects of "rest" and "motion" of bodies have led to the division of the subject into two parts:

(1) **Statics,** which deals with bodies at rest.
(2) **Dynamics,** which is concerned with bodies in motion.

There is a further branch of the subject, which frequently has a volume to itself, viz. **Hydrostatics,** in which are studied the application to liquids and gases of those principles which have been examined for solids.

It will be noticed that the word "body" has been employed above, and it will continually enter into the work of this volume. The words "body" and "object" are meant to include all things which, on earth, have weight, a term which we shall examine more closely later. We assume, for the present, that all such "bodies" are "rigid—i.e. different parts of them always retain the same relative positions.

3. Forces

In both statics and dynamics we are concerned with forces. A single force acting on a body will cause it to start moving, first slowly, and then, if the force continues to be applied, faster and faster.

A single force on a body causes it to accelerate.

If we see a force acting on a body, e.g. a tug-of-war team pulling on a rope, and yet the body does not move, we deduce that there is an equal force pulling in the opposite direction.

If the body is at rest while several forces are acting on it the forces are said to be in equilibrium.

In the simple example above the forces from the two teams were in equilibrium.

A third situation is one in which a body is not accelerating, but neither is it at rest, e.g. a sledge being pulled along at a steady speed. The forces acting on the sledge are:

(1) Those trying to speed it up, e.g. the pull of the rope.

(2) Those trying to slow it down, e.g. the friction on the runners.

If the sledge continues at a steady speed these two groups of forces must be equal. Otherwise it would either speed up or slow down.

A force may have many different origins. It could be muscular or from gravity. Electrical and magnetic effects could produce it, as could a wind or even friction. The ways in which forces cause bodies to move will be considered further in Chapter X.

4. Transmission of force. Tension

The rope in the tug-of-war referred to above is used to make one team feel the force of the other. The rope is subject to a tension. A string or rope is a common means of transmitting a force and in most cases the tension is the same throughout it.

5. Weight

If a stone is dropped it accelerates towards the earth. Clearly a force must have caused this—a force of gravity. We feel this force when we hold the stone in our hands. It acts vertically downwards and we call it the weight, W, of the stone. If the stone is not moving there must be an equal and opposite force, F, upwards, provided, perhaps, by muscular effort.

This is represented in Fig. 1.

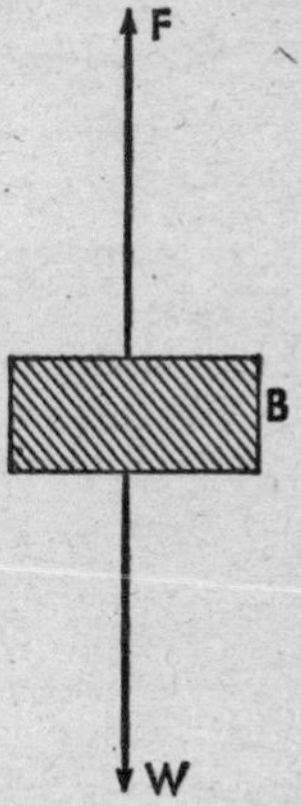

Fig. 1.

If $F = W$ the body is at rest.
If $F > W$ the body accelerates upwards.
If $F < W$ the body accelerates downwards.

6. Mass

It is common experience that if we apply the same force to a very heavy body and then to a lighter body the lighter body accelerates more than the heavy one. We say that the heavy body has more inertia. We are also familiar with pictures of astronauts floating in space. If they are far away from the earth's gravitational attraction they have no weight. Yet under these conditions the body which was heavy on earth would still be harder to accelerate than the one which was light. *The inertia does not change.* **The property which measures inertia is called mass.** It is sometimes described as the amount of matter in a body.

The *mass* of a body *does not* depend on where the body is situated.

The *weight* of a body *does* depend on where the body is situated.

It is worth noting that when we buy, for example, potatoes the price quoted should be for a given mass.

7. Measuring mass and force

Mass is measured in kilogrammes. One kilogramme, kg, is the mass of a standard lump of metal (platinum) kept near Paris.

The accepted unit of force is the Newton, but we shall defer using this unit until after Chapter X.

We have seen that the weight of a body is the force with which the earth attracts it. This gives us an alternative unit of force, the "kilogramme force", kgf.

1 *kgf is the force with which the earth attracts a body which has a mass of* 1 *kg.*

Strictly speaking we should specify the geographical position of the body, for there is a slight difference in the earth's attraction at different places.

8. A spring balance

The effect on our muscles cannot be used to distinguish accurately between forces of different magnitudes. We could detect that a stone weighing 2 kgf was heavier than one weighing 1 kgf, but we could not say by how much. We therefore require an instrument which can measure forces more exactly. It should be able to measure hori-

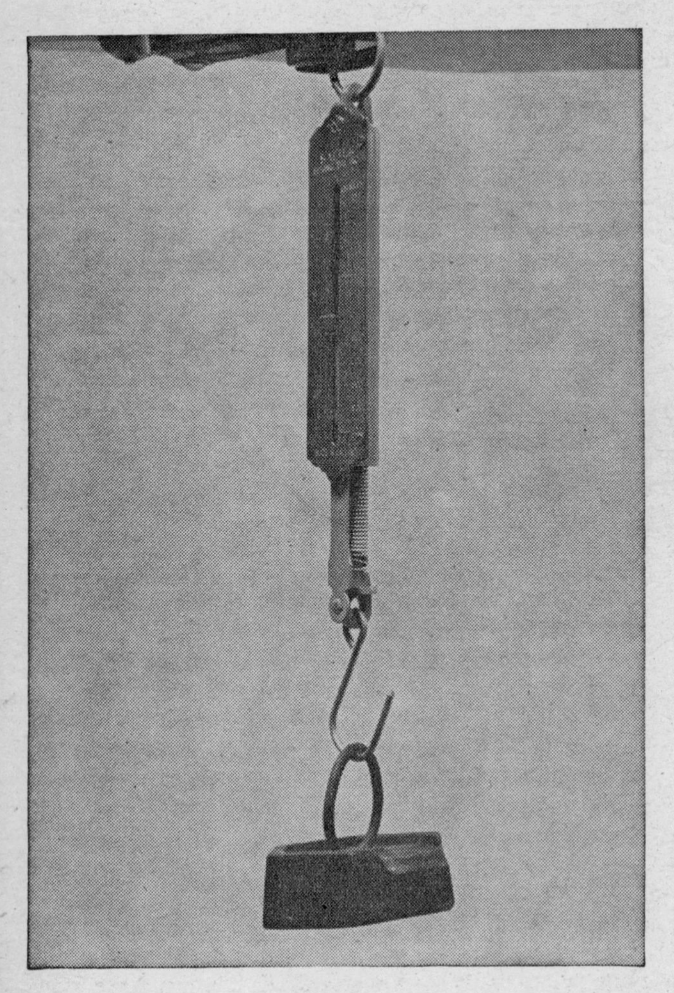

FIG. 2.

zontal forces (e.g. the force on the tug-of-war rope), or inclined forces, as well as vertical ones. Many instruments will do this, but one of the simplest is a spring balance, an example of which is shown in Fig. 2. If you take a coiled spring and attach an object weighing 1 kgf, it will extend in length by a certain amount. Now, it can be demonstrated by experiment that the extension in the length of the spring is proportional to the magnitude of the force. If, therefore, an object weighing 2 kgf be attached to the spring, the extension in its length will be twice the extension for that weighing 1 kgf.

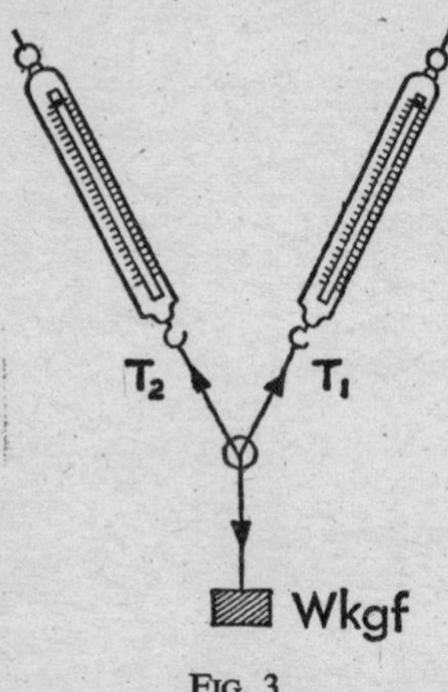

FIG. 3.

In this way it is possible to construct a scale by the side of the spring on which will be shown the magnitude of the force which produces a certain extension. Quite clearly it would be equally easy to make such a balance which measured the forces in Newtons.

The spring balance can be used to measure force acting in any direction. Thus, if a loading weighing W kgf is supported by two spring balances as shown in Fig. 3, then the tension in each string, shown as T_1 and T_2 in the figure, can be measured on the corresponding spring balance.

9. Standard Units and abbreviations

The speed of a person is the distance he travels in a given length of time. If the distance is measured in metres (m) and the time is measured in seconds (s) the speed, which is distance ÷ time, is measured in "metres per second" (m/s).

A "metre per second" is known as a "derived unit", because it measures one quantity, speed, in terms of other quantities, length and time. Sometimes derived units are given a special name of their own, e.g. a "nautical mile per hour" is called a "knot". With the exception of the kgf the units used in this book are in accordance with the recommendations of the British Standards Institution

Table I
Standard units and their abbreviations

Quantity	Name of unit	Standard abbreviation	Relationship to other units
Time	second	s	—
	minute	min	$60\ s = 1\ min$
	hour	h	$3600\ s = 1\ h$
Length	metre	m	—
Volume	litre	l	$10^{-3}\ m^3 = 1\ l$
Speed	knot*	kn*	$0{\cdot}514\ m/s = 1\ kn$
Mass	kilogramme	kg	—
	tonne*	t*	$1000\ kg = 1\ t$
Force	newton	N	$1\ kg\ m/s^2 = 1\ N$
Pressure	bar*	bar*	$10^{-5}\ N/m^2 = 1\ bar$
Energy and Work	joule	J	$1\ N\ m = 1\ J$
Power	watt	W	$1\ J/s = 1\ W$

publication PD 5686, and those units relevant to this book are shown in Table I. Those which are marked with an asterisk are used only rarely.

10. Multiples and sub-multiples of units

Two thousandths of a metre could be written as

$$\frac{2}{1000}\ m \quad \text{or} \quad 0{\cdot}002\ m \quad \text{or} \quad 2 \times 10^{-3}\ m.$$

Twenty thousand metres could be written as

$$20\,000\ m \quad \text{or} \quad 2 \times 10^4\ m.$$

The index, in this case "4", tells us how many times we must multiply the first number by ten. 10^{-3} tells us that we must divide by 10^3, i.e. we must divide by 1000.

To avoid having many factors of 10 apparent when the

value of a quantity is quoted, a standard list of prefixes has been agreed to. This is shown in Table II.

Using this table we find that 0·002 m becomes 2 mm, and that 20 000 N becomes 20 kN.

Table II

List of standard prefixes

Factor by which the unit is multiplied	Prefix	Symbol
10^{12}	tera	T
10^{9}	giga	G
10^{6}	mega	M
10^{3}	kilo	k
10^{2}	hecto	h
10	deca	da
10^{-1}	deci	d
10^{-2}	centi	c
10^{-3}	milli	m
10^{-6}	micro	μ
10^{-9}	nano	n
10^{-12}	pico	p

In the few places where ambiguity might arise it is wise to write the name of the unit in full rather than use the abbreviation. Thus

2 m multiplied by 3 N = 6 m N or 6 metre Newtons.
0·006 N = 6 mN or 6 milliNewtons.

CHAPTER II

THE LEVER

11. Machines

A machine is a contrivance by means of which a force applied at one part of the machine is transmitted to another in order to secure an advantage for some particular purpose.

For example, suppose I wish to raise a heavy stone indicated in Fig. 4 by *L*. I insert an iron bar under one

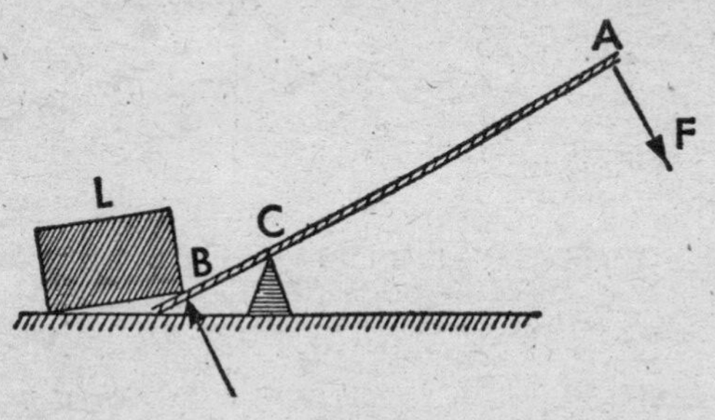

FIG. 4.

edge at *B* and pivot it on a suitable object, such as a stone, at *C*. By the application of a comparatively small force, *F*, at *A*, this force is transmitted to *B*, acts upwards, and raises the stone.

The bar is thus a contrivance by means of which it is possible to transmit an applied force to secure the advantage of a larger force acting at another point.

The rod is a simple example of a **lever,** which is probably one of the earliest machines to be used by man, and today, in various forms and combinations, is the most widely used of all machines.

We must now proceed to examine the principles underlying it, for it is only when these are understood that progress may be made in developing the uses of the machine.

12. The principles of the lever

The student can easily discover the mechanical principles of a lever by a few simple experiments.

A long bar is the essential thing for a lever. We will therefore begin by experimenting with a long rod or bar which is graduated in inches or centimetres, in order to facilitate the experiments. A metre rule is very suitable. It will be necessary to have some means of pivoting the rod, so that it may turn about the point of the pivoting. This could be done by boring holes, with smooth interior surfaces, then passing a knitting-needle, or something similar, through a hole. The needle is then supported in some way.

We will begin by pivoting the stick in the middle. Let *AB* (Fig. 5) represent such a bar and *C* the centre

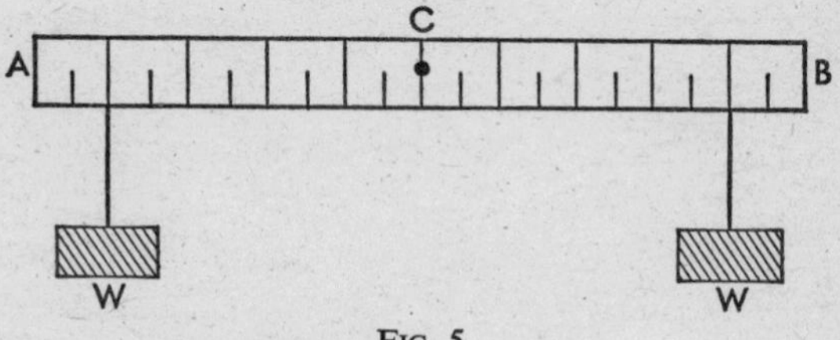

FIG. 5.

hole about which it is pivoted. If this is done accurately the bar will rest in a horizontal position. In this form the lever is a simple balance, and if equal weights, *W*, are suspended at equal distances from the centre, the rod will be in equilibrium. It should be noted that the weight near *B* **tends to turn the rod** about *C* **in a clockwise direction,** while that near *A* **tends to turn the rod in an anti-clockwise direction.**

The turning effects of these two weights, each of which is a force acting vertically downwards, balance one another and so the equilibrium is not disturbed.

13. Turning moments

The effect of hanging a weight at any point on one arm of the lever is to cause the lever to turn about the point of support. We must now therefore investigate how this turning effect is affected by:

(1) The **magnitude** of the weight hung on.

(2) The **distance** of the weight from the pivot, called the **fulcrum.**

(3) The **position of the fulcrum.**

First let us consider experiments in which we vary the magnitudes of the weights and their distance from the fulcrum, which we will still keep at the centre of the lever.

Let a weight of 2 kgf be suspended at *D* (Fig. 6) 40 cm

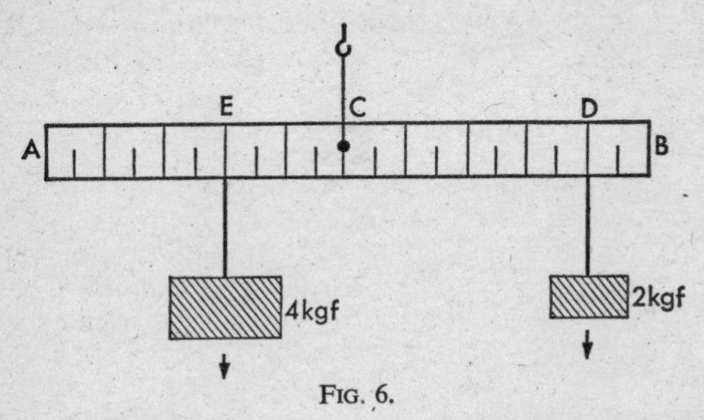

FIG. 6.

from *C*, the fulcrum. Let this be balanced by 4 kgf placed on the other arm. It will be found that to balance the 2 kgf we must place the 4 kgf at *E*, 20 cm from *C*.

Thus 2 kgf acting 40 cm from *C* on one arm
balances 4 kgf acting 20 cm from *C* on the other arm.

In other words—

The turning effect in a clockwise direction of 2 kgf acting 40 cm from *C* balances the turning effect in an anti-clockwise direction of 4 kgf acting 20 cm from *C*.

It will be noticed that the products of weight and

distance in the two cases, viz. 2×40 and 4×20, are equal.

This product of force (or weight) and distance is called the turning moment of the force.

We further notice in this case that when the turning moments balance about the fulcrum, they are equal. Similar experiments in which the weights and distance are varied lead to the conclusion that this is true generally.

In these experiments the fulcrum or pivot is at the centre of the lever.

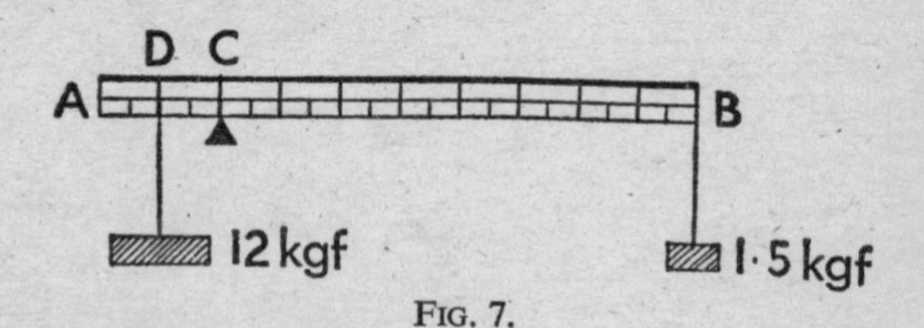

FIG. 7.

We must now try similar experiments when the fulcrum is in any position. This time instead of suspending the bar it is balanced about a knife-edge at *C* (the fulcrum).

Let the lever, *AB* (Fig. 7), be 100 cm long. Let the fulcrum, *C*, be 20 cm from *A*.

A weight of 12 kgf is hung at *D*, 10 cm from *C*.

If we experiment to find what weight placed at *B*, the end of the lever, will balance this, it will be found that 1·5 kgf is needed.

Thus 12 kgf at 10 cm from *C* is balanced by
1·5 kgf at 80 cm from *C*.

The turning moment (clockwise) of

$$1{\cdot}5 \text{ kgf at } B = 1{\cdot}5 \times 80 = 120 \text{ kgf cm.}$$

The turning moment (anti-clockwise) of

$$12 \text{ kgf at } D = 12 \times 10 = 120 \text{ kgf cm.}$$

The turning moments of the two weights are equal and opposite.

Similar experiments with varying weights and distances and different positions of the fulcrum lead to the same conclusion, viz.—

When the turning moments of two forces about the fulcrum of a lever are equal and opposite, the lever is in equilibrium.

The converse is also true.

Note. When turning moments are thus calculated with regard to the fulcrum at *C*, we say that "we take moments (or turning moments) about *C*".

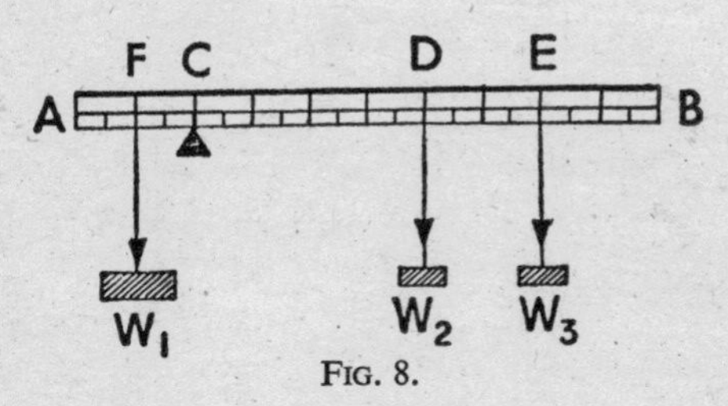

FIG. 8.

In the previous experiments, and the conclusions drawn from them, we have considered the case in which one weight only was placed on each arm. To complete the investigation experiments should be made to discover what happens if more than one weight is acting.

We can proceed as follows:

Let a bar, *AB*, as before, rest on a fulcrum at *C* (Fig. 8).

Let three weights, W_1, W_2, W_3, be suspended from the bar at *F*, *D* and *E*. Let d_1, d_2, d_3 be their distances from the fulcrum.

Let these weights be so arranged that the bar is in horizontal equilibrium—i.e. it balances about *C* and remains horizontal.

Now W_2 **and** W_3 **exert a clockwise turning movement** about *C*.

While W_1 **exerts an anti-clockwise effect.**

If there is equilibrium these must balance—i.e. the clockwise turning moment must equal the anti-clockwise moment.

$$\therefore \quad (W_1 \times d_1) = (W_2 \times d_2) + (W_3 \times d_3).$$

This should be verified experimentally by actual weights and measured distances.

14. Relation between weights and distances

Let AB be a bar with the fulcrum, C, at its centre (Fig. 9).

Let W_1 and W_2 be unequal weights hung on the bar so that the bar rests in horizontal equilibrium.

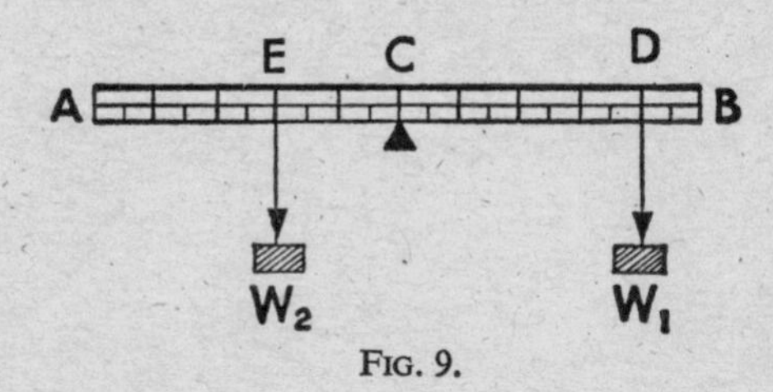

FIG. 9.

Let d_1 and d_2 be the respective distances from C.

Since there is equilibrium, the clockwise turning moments equal the anti-clockwise.

$$\therefore \qquad W_1 \times d_1 = W_2 \times d_2.$$

This can be written in the form:

$$\frac{W_1}{W_2} = \frac{d_2}{d_1}.$$

15. The weight of the lever

We have hitherto disregarded the weight of the lever, assuming the bar to be "light", so that its weight does not materially affect the conclusions. But if the lever used is not light, discrepancies will have appeared in the results of the experiments.

It must first be noted that the weight will, of course, act vertically downwards. Early mathematicians assumed it as axiomatic that **the whole of the weight could be considered as acting at the centre of the lever,** provided that the lever is uniform. This is confirmed by the fact that it can be supported in horizontal equilibrium at its centre. If it be thus suspended by means of a spring balance it will be found that the balance registers the whole weight of the lever, which must therefore act at the point of suspension.

The following experiment will serve to demonstrate this, as well as to illustrate our previous conclusions.

Suspend the lever from a point, *O* (*see* Fig. 10).

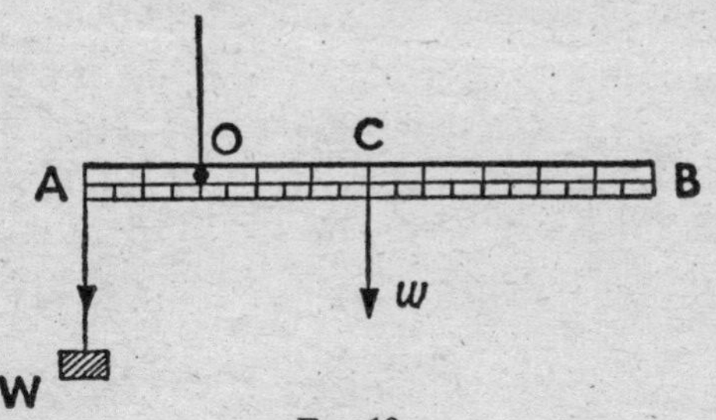

FIG. 10.

Suspend also a weight, *W*, at *A*, of such a magnitude that the rod rests in horizontal equilibrium.

Then, from our previous conclusions, the anti-clockwise movement of *W* acting at *A* must be counterbalanced by the weight of the lever, which, as the moment must be clockwise, must act somewhere along *OB*.

Let the point at which it acts be x cm from *O*.

Then by the conclusions of § 13,

the moment of $W =$ moment of the weight of the lever.

Let this weight be w.

Then $$w \times x = W \times OA.$$

$$\therefore \qquad x = \frac{W \times OA}{w}.$$

Now *OA* can be measured and W and w are known.

$\therefore$ x can be calculated, and so we find the distance from *O* at which the weight must act.

From the experiment you will find that $x = OC$, where *C* is the centre of the lever.

Thus the weight of the lever can be considered as acting at the centre of the lever.

The point at which the force of gravity on the lever—i.e. its weight—acts, is called the **Centre of Gravity** of the rod. **If the lever is not uniform the centre of gravity will not be at the centre of the lever,** but its position can be found by using the above method.

The term "Centre of Gravity" is introduced now as it arises naturally out of the experiments performed, but the full consideration of it is postponed until Chapter III.

16. The effect of inclined forces

In the examples considered so far the forces (weights) have all acted vertically downwards, and so have been perpendicular to the bar. If the force is not perpendicular to the bar it can be shown experimentally that the distance measured must be from the fulcrum to the line of action of the force. This is illustrated in Fig. 11, in which a

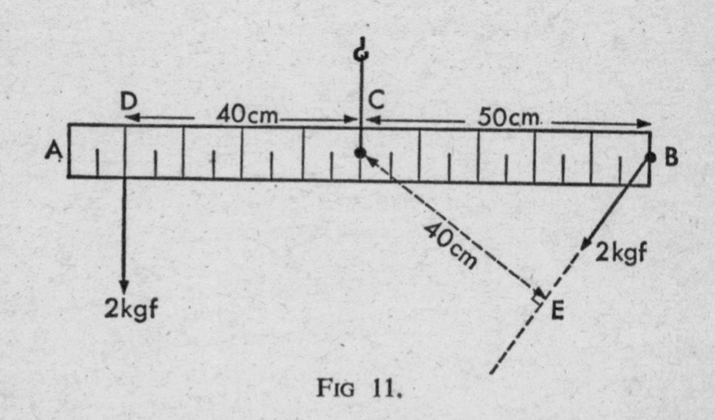

FIG 11.

force of 2 kgf, at *D*, 40 cm from the fulcrum, is sufficient to balance a force of 2 kgf acting on the bar at *B*, 50 cm from the fulcrum. This second force is not vertical, and the distance from the fulcrum to its line of action is again 40 cm (*CE*).

The turning moment of a force about a point is the product of the force and the distance from the point to the line of action of the force.

Exercise 1

1. A light rod, AB, 0·30 m long, is pivoted at its centre C, and a load of 8 kgf is hung 0·12 m from C.

(*a*) Where on CB should a load of 15 kgf be placed to balance this?
(*b*) What load should be placed 0·1 m from C to preserve equilibrium?

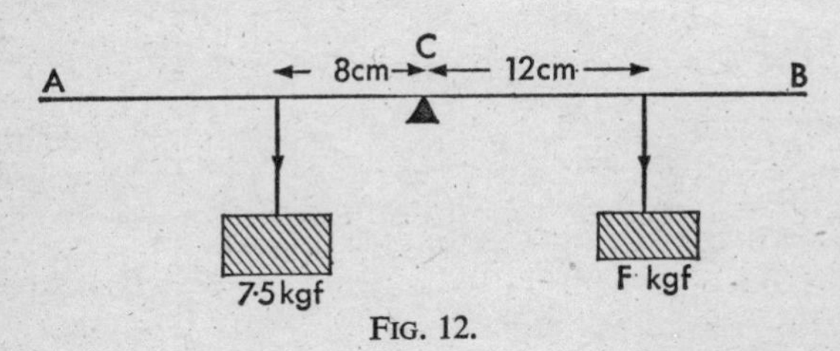

FIG. 12.

2. Fig. 12 shows a light rod, AB, of length 0·36 m pivoted at its centre C. Loads are hung as indicated.

(*a*) What is the value of F?
(*b*) If the 7·5 kgf were acting at A, what would the value of F then be?
(*c*) If a weight of 5 kgf were hung at B, what weight must be placed 0·12 m from A to preserve equilibrium ($F = 0$)?

3. A light rod 0·40 m long has a fulcrum at its centre. Loads of 4·5 kgf and 2·5 kgf are attached on one arm at distances 0·08 m and 0·05 m respectively from the fulcrum. Where must a single load of 4 kgf be placed on the other arm to preserve equilibrium?

4. A light rod 4 m long balances about a fulcrum 125 cm from one end. A load of 18 gf is placed on the shorter arm 60 cm from the fulcrum.

(*a*) What weight placed 250 cm from the fulcrum on the long arm will balance the loaded rod?
(*b*) Where must a load of 5 gf be placed for the same purpose?

5. A uniform rod 0·96 m long is supported on a fulcrum 0·32 m from one end. A load of 7 gf placed at the end of the shorter arm causes the rod to rest in horizontal equilibrium. What is the weight of the rod?

6. A uniform rod 36 cm long and weighing 4 gf is supported on a fulcrum 5 cm from one end. At this end a load of 0·192 kgf is attached. What load placed on the other arm, 25 cm from the fulcrum, will balance this weight?

7. It is desired to raise a mass of 300 kg placed at the end of a horizontal uniform iron bar, 3 m long and weighing 11 kgf. To effect this a fulcrum is placed 25 cm from this end and the bar pivoted on it. What effort must be exerted at the other end of the bar so that the weight can just be moved?

8. A uniform bar 6 m long, and of unknown weight, rests on a fulcrum 1 m from one end. It is in equilibrium when loads of 6 kgf and 36 kgf are hung at the end of the bar.

(*a*) What is the weight of the bar?

(*b*) What load at the end of the long arm will balance a load of 24 kgf at the end of the short arm?

9. A uniform bar, 2·4 m long and of weight 5 kgf rests on a fulcrum 0·4 m from one end. On the other arm loads of 3 kgf each are hung at every 0·5 m from the fulcrum. What load at the end of the short arm will preserve horizontal equilibrium?

10. A heavy uniform bar is 30 cm long and weighs 2 kgf. A weight of 3 kgf is placed at one end, and at the other a weight of 5 kgf. At what point on the bar must it be supported so that it balances and does not turn?

17. Force on the fulcrum. Resultant force

In the previous experiments weights have been hung on a rod which was supported at the fulcrum; consequently there is a force on the fulcrum. To ascertain the amount of this force the following experiment may be performed.

Let a bar (Fig. 13) be loaded with two weights, W_1

and W_2, placed so that there is equilibrium about a fulcrum, C. Replace the fulcrum by a hook attached to a spring balance and let this be raised slightly so that it takes the weight of the whole system. If there is still equilibrium, then the spring balance will register the force which was formerly borne by the fulcrum.

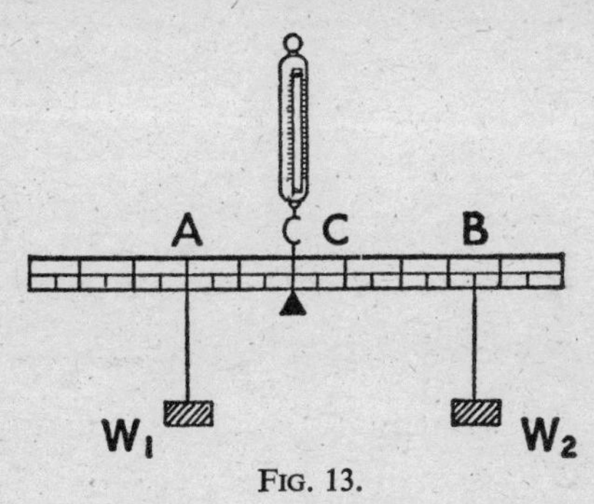

FIG. 13.

Taking the reading of the spring balance, you will find, as you probably expected, that it is equal to the **sum of the weights**—i.e. to $W_1 + W_2$. If the rod is not a light one the **weight of the lever** will also be borne by the fulcrum, or the spring balance which replaces it.

Let **R be the force recorded.**
Then $\mathbf{R} = W_1 + W_2$.

If other weights be placed on the lever, equilibrium being preserved, the force on the fulcrum is increased by these.

It is clear that if the two weights or forces, W_1 and W_2, were replaced at C by a single weight or force, R, equal to their sum, the spring balance would register the same amount as before; the effect on the fulcrum, C, would be the same.

The single force which thus replaces two separate forces acting on a body, and has the same effect on the body, is called the resultant of the forces.

It should be noted that the forces represented by W_1 and W_2, being due to gravity, are parallel forces.

The following points should be noted about the **resultant of parallel forces:**

(1) **It is parallel** to the forces it can replace and its **direction is the same.**

(2) **It is equal to the sum of these forces.**

(3) The **turning moments of the forces** represented by W_1 and W_2 are $W_1 \times AC$ and $W_2 \times BC$.

These are equal, since there is equilibrium. Therefore, **the moments of the forces about C, the point at which the resultant acts, are equal.**

(4) Since $W_1 \times AC = W_2 \times BC$

$$\therefore \quad \frac{AC}{BC} = \frac{W_2}{W_1}.$$

$\therefore$ **the point through which the resultant of the two parallel forces acts divides the distance between them in the inverse ratio of their magnitudes.**

18. Centre of force

The converse of the above is also true—viz. if two parallel forces represented by W_1 and W_2 act on a bar

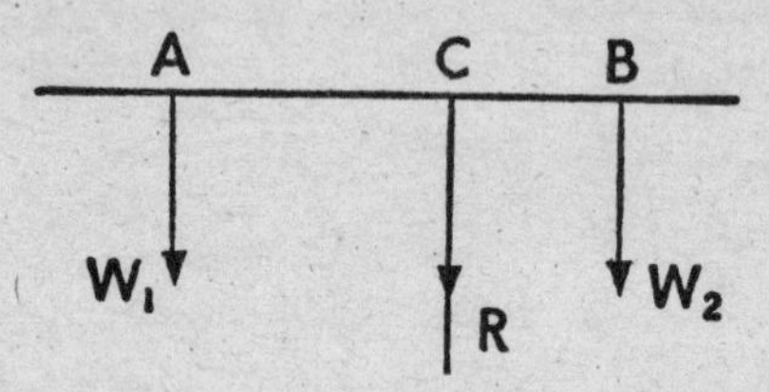

FIG. 14.

at A and B (*see* Fig. 14), it is possible to find a point, C, between them so that

$$\frac{AC}{CB} = \frac{W_2}{W_1},$$

and

$$W_1 \times AC = W_2 \times CB.$$

But these are the moments of W_1 and W_2 about C.

$\therefore$ it is possible to find a point so that the moments about it of the two forces are equal and opposite and there is therefore equilibrium.

Such a point is called **the centre of force,** and through this point the resultant acts.

19. Bars resting on two supports

The principles that have been established in the preceding pages may be extended to the case of loaded rods, bars or beams which rest on two supports. In problems arising from this arrangement it is important to be able to calculate the thrust of the supports on the bar or beam, or, conversely, of the thrusts on the supports. The practical applications are very important. The following is a description of an experiment by means of which these thrusts may be found practically.

In Fig. 15, *AB* represents a heavy uniform bar, or beam, resting on two supports at *D* and *E*. These may be attached to compression balances, or the bar may be suspended from spring balances fixed at these points, as was shown in § 17.

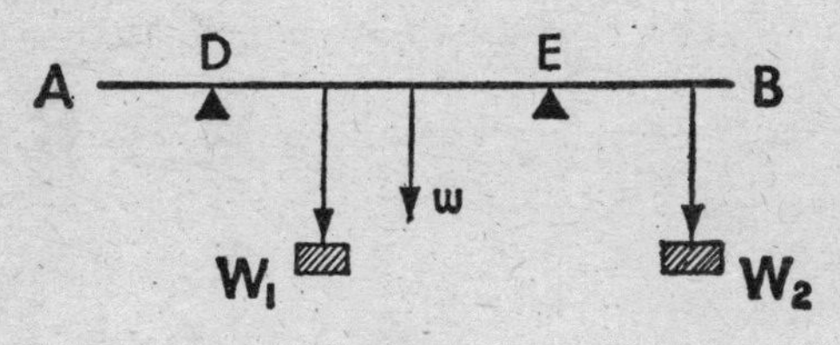

FIG. 15.

Let loads W_1 and W_2 be suspended from two points.

Let w be the weight of the bar, acting at the centre of it.

If the recorded forces registered at *D* and *E* are examined it will be found that—

$$\text{total forces at } D \text{ and } E = W_1 + W_2 + w.$$

This was to be expected, since these two supports must take the whole of the downward forces, due to gravity. It will be seen, however, from the readings of the two balances that this force is **not divided evenly** between the two supports. Our problem is to discover how, in any given instance, these may be calculated.

We will first consider the simple case of a heavy bar, in which the centre of gravity is not at the centre of the bar and no additional loads are placed on it.

(*a*) **Experimental method**

Let *AB* (Fig. 16) be a heavy bar 24 cm long, weight 6 kgf, and supported at *A* and *B*.

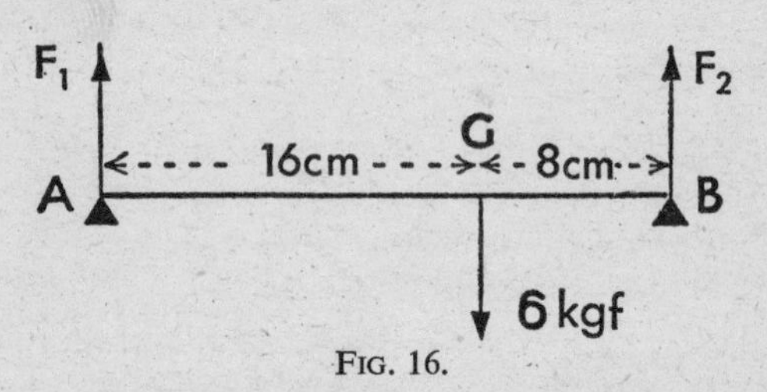

FIG. 16.

The centre of gravity is not at the centre of the rod but known to be at *G*, where $AG = 16$ cm, $GB = 8$ cm.

Let F_1, F_2 be the forces at *A* and *B*.

Consider the forces acting on the bar *AB*. These are:

downward—the weight of the bar, 6 kgf;
upward—$F_1 + F_2$ (equal and opposite to thrust on the supports).

Since there is equilibrium, these must be equal.

$$\therefore \qquad F_1 + F_2 = 6.$$

Reading the balances at *A* and *B* we find:

$$F_2 = 4, F_1 = 2,$$

confirming the conclusion already reached that

$$F_1 + F_2 = 6.$$

We note that

$$F_2 : F_1 = 2:1 = 16:8 = AG:GB.$$

Thus we find from the experiment that **the forces on the supports are inversely proportional to the distances from *G*.**

This can be confirmed by similar experiments.

(*b*) **Use of principles of moments**

Since F_1 and F_2 represent the forces of the bar on *A* and *B*, equal and opposite thrusts must be exerted on the bar, as is stated above, since there is equilibrium at these points.

Suppose the support at *A* were removed. The bar would **turn about the other support at *B*,** and we could regard the bar as a lever, just on the point of turning about *B*.

The bar is then subject to the turning moments about *B* of:

(1) The weight at *G*—**anti-clockwise.**
(2) F_1 at *A*—**clockwise.**

The thrust F_2 at *B* has no effect on rotation about *B*.

These turning moments must be equal.

$$\therefore \quad F_1 \times 24 = 6 \times 8.$$
$$\therefore \quad \mathbf{F_1 = 2\ kgf.}$$

Similarly, if we imagine the bar on the point of turning about *A* the turning moments are:

(1) $F_2 \times 24$—**anti-clockwise**
(2) 6×16—**clockwise**

$$\therefore \quad F_2 \times 24 = 6 \times 16$$
$$\text{and} \quad \mathbf{F_2 = 4\ kgf.}$$

The methods adopted in this case are also applicable when loads are attached to the bar. This is illustrated in the following example:

20. Worked example

A uniform heavy bar, AB, of length 120 *cm and weighing* 5 *kgf, is supported at one end A and also at C, which is* 20 *cm from B. A load of* 10 *kgf is suspended at D, which is* 20 *cm from C. Find the forces on the supports.*

Fig. 17 represents the rod *AB*.

As the rod is uniform, the centre of gravity is at the centre *G*, of the rod.

Let F_1 = upward thrust at *A*
and F_2 = upward thrust at *C*.

These are equal and opposite to the forces from the bar on the supports at *A* and *C*.

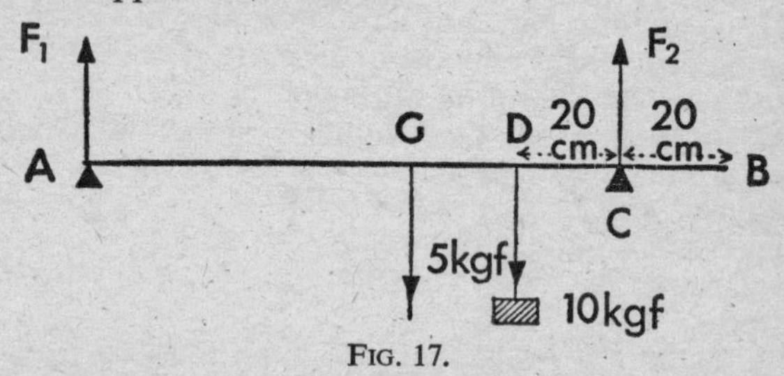

FIG. 17.

Forces acting on the bar:

(1) Weight, 5 kgf, downward at *G*.
(2) Load, 10 kgf, downward at *D*.
(3) F_1 upward at *A*.
(4) F_2 upward at *C*.

Now, suppose that the support at *A* were to be removed so that the bar begins to rotate about *C*. We find the turning moments of the forces which tend to make the bar rotate about *C*. They are:

$F_1 \times 100$—**clockwise.**
$(5 \times 40) + (10 \times 20)$—**anti-clockwise.**

As there is equilibrium about *C*, these must be equal.

$$\therefore \quad F_1 \times 100 = (5 \times 40) + (10 \times 20)$$
$$100F_1 = 400$$
$$\therefore \quad F_1 = \mathbf{4\ kgf.}$$

Similarly, equating moments about *A*

Clockwise **Anti-Clockwise**

$$F_2 \times 100 = (5 \times 60) + (10 \times 80)$$
$$\therefore \quad 100F_2 = 1100$$
$$\therefore \quad F_2 = \mathbf{11\ kgf.}$$

We may check by noting that since there is equilibrium the forces acting vertically upwards must be equal to those acting vertically downwards. We have:

upward $F_1 + F_2 = 4 + 11 = 15$ kgf,
downward 5 kgf + 10 kgf = 15 kgf.

Exercise 2

1. A uniform bar, 2 m long, weighing 10 kgf, is supported at one end and also 0·25 m from the other end. Find the load carried on the supports.
2. A mass of 60 kg is hung on a uniform wooden pole 1·6 m long and weighing 2 kgf. It is then carried by two men, one at each end of the pole. It is so arranged that the position of the weight on the pole is 0·5 m from the stronger man. What weight will be borne by each of the men?
3. A uniform beam 1·2 m long and weighing 20 kgf is supported at each end. Loads of 50 kgf and 60 kgf are carried at distances of 0·2 m and 0·8 m respectively from one end. Find the load on each support.

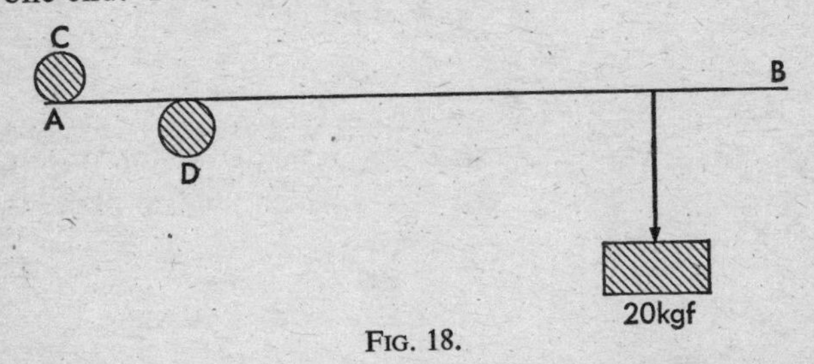

FIG. 18.

4. A uniform plank 2 m long and weighing 10 kgf is supported between two steel pegs, *C* and *D*, arranged as shown in Fig. 18; one, *C*, at one end of the plank, and the other, *D*, 35 cm from the end. The plank carries a load of 20 kgf suspended 35 cm from the other end, *B*. Find the thrust on each of the pegs.
5. A uniform wooden plank weighing 20 kgf and 4 m long is placed symmetrically on two supports 3 m apart. A load of 80 kgf is attached to one end and 100 kgf to the other. What is the thrust on the supports?

6. A uniform bar 2 m long weighs 2 kgf and is supported at its ends. A 7 kg mass is hung from the bar 0·5 m from one end, and a 4 kg mass hung 0·75 m from the other end. Find the thrust on the two supports.

7. A heavy uniform beam, AB, weighing 1·5 tf is supported at A and at a point C, one quarter the length of the beam from B. A load of 0·5 tf rests on the end B. What are the forces acting on the supports A and C?

8. A plank 4 m long of mass 30 kg rests on supports at each end. A man weighing 75 kgf stands on the plank 1 m from one end. What is the force on each of the ends?

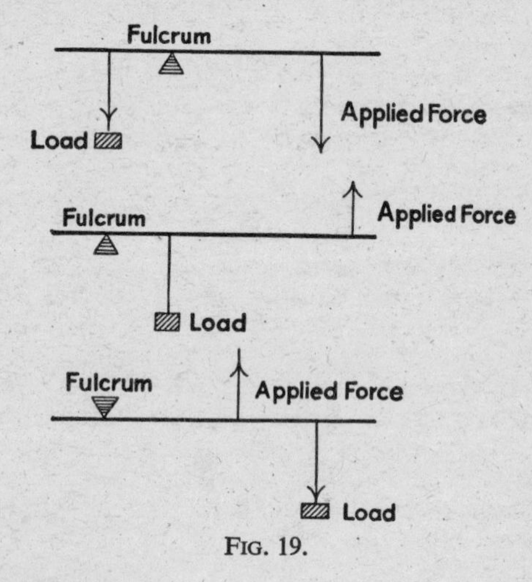

FIG. 19.

21. Orders of levers

As we have seen, a lever, in addition to a bar or rod, involves:

(1) A load. (2) A fulcrum. (3) An applied force.

The relative positions which these may occupy on the lever may vary. Hitherto, in the examples which we have examined, the **position of the fulcrum has been between the load and the applied force.** As the positions of these are altered we have different types of levers. Three arrangements are possible, and these, since the time of Archimedes, have been called the **three orders of levers.**

The relative positions in these orders are shown in Fig. 19.

22. The principle of moments in the three orders

The principles established for levers of the first order apply to the other orders and may be verified experimentally by the student. In particular the principle of moments is very important, and it is worth while considering its application in the three orders. The principle was:

When the lever is in equilibrium the turning moments of the forces which tend to produce motion in a clockwise direction about the fulcrum are equal to those of the forces tending to produce motion in an anti-clockwise direction.

In applying this principle to the three orders it may be noted that:

(1) In the **first order,** since the load and applied force act on opposite sides of the fulcrum, one will naturally be clockwise and the other anti-clockwise, though they both **act in the same direction.**

(2) In the **second and third orders,** if one is to be clockwise and the other anti-clockwise, they **must act in the opposite directions.** On examining Fig. 19 it will be seen that in levers of the second and third orders the **moment of the load about the fulcrum is clockwise, while the moment of the applied force is anti-clockwise.**

23. Relative advantages of the three orders

It will be noticed that in the first and second orders the length of the "applied force" arm is, in general, greater than that of the "load" arm, but in the third order the converse holds.

Consequently, in the first and second orders there is an advantage, since the load moved is greater than the force applied. In the third order the applied force is greater than the load.

On the other hand, in the first two orders the applied force has to move through a greater distance than the load, while in the third order the applied force moves through a shorter distance. Many of the movements in the human body are made by muscular action which is applied as in the third order of levers. Here it is an advantage that the applied force should move through a short distance.

24. Practical examples of levers

The following are a few examples of the practical applications of levers in daily life.

First order

A balance, crowbar (with fulcrum arranged as in Fig. 20(*a*)), a poker (pivoted on a bar), a pair of scissors

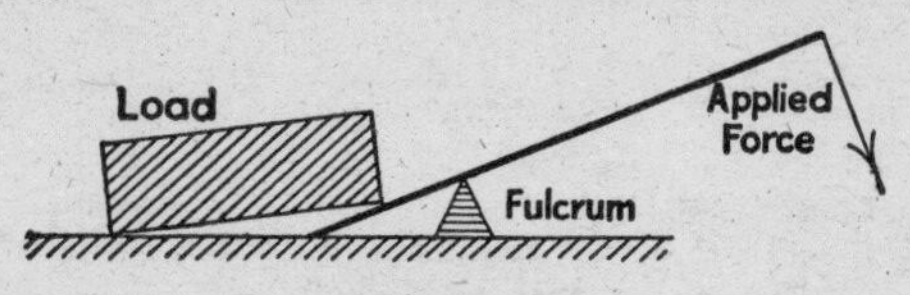

FIG. 20(*a*).

or pliers (double lever hinged), steel-yard, pump handle (Fig. 20(*b*)).

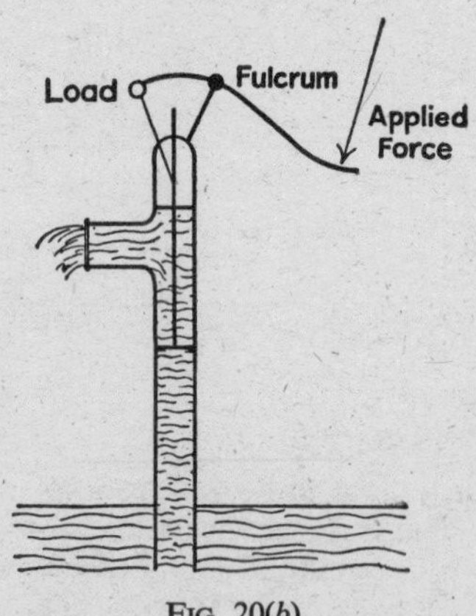

FIG. 20(*b*).

Second order

A crowbar (with fulcrum on ground as in Fig. 21(*a*)), a wheelbarrow, pair of nutcrackers (double-hinged), oar of a boat, a safety valve lever (Fig. 21 (*b*)).

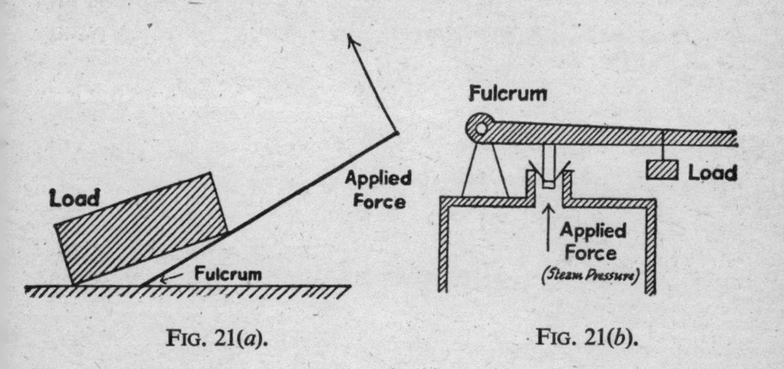

FIG. 21(*a*). FIG. 21(*b*).

Third order

The forearm of the human body (Fig. 22(*a*)), a pair of sugar-tongs (double lever) (Fig. 22(*b*)).

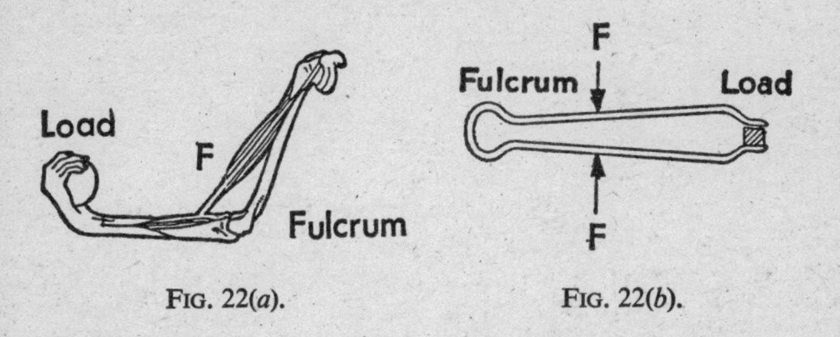

FIG. 22(*a*). FIG. 22(*b*).

25. Combinations of levers

Combinations of levers are frequent in complicated mechanisms. The striking mechanism of a typewriter is an example of a combination of levers of the first and second orders (*see* Fig. 23). When a note is struck on a piano, a combination of levers of all three orders is employed, for the purpose of transmitting the action to the wires. In an aneroid barometer a combination of levers is also employed.

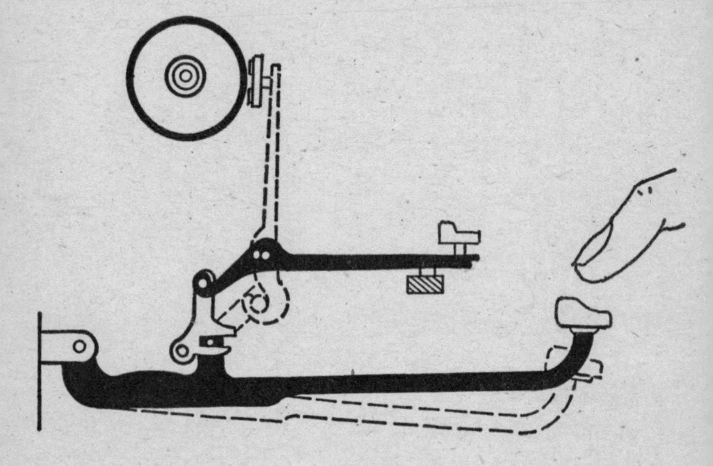

FIG. 23.

26. A simple pulley

The pulley is another very useful machine. In its simplest form it consists of a wheel with outer rim grooved to permit of a rope travelling round it (Fig. 24). The axle of the wheel is attached to a fixed beam or other support. In this form the pulley is employed merely to change the direction of an applied force.

Thus in Fig. 24 it will be seen that a weight, W, can be pulled *upwards* by a force, F, acting *downwards* on the rope which is attached to W and passes round the pulley.

If the pulley is assumed to be "smooth"—i.e. there is no friction—then the tension in the rope is the same throughout and $F = W$.

It should be noted that, since there is a pull, F, in the rope on each side of the pulley, the total force acting on the pulley is $2F$.

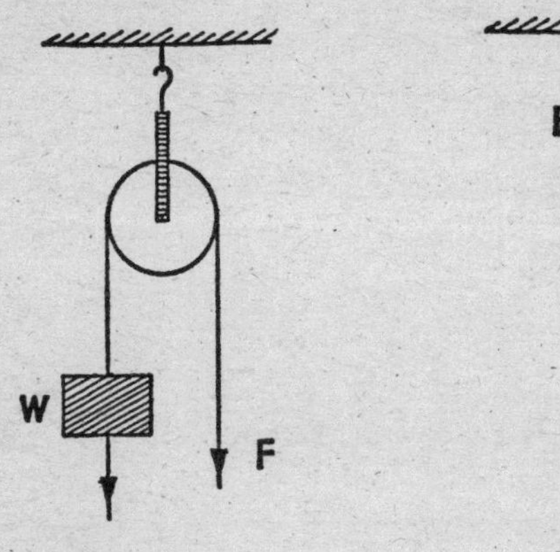

FIG. 24.

FIG. 25.

A movable pulley

In the arrangement shown in Fig. 25 the pulley is a movable one. A rope is fixed to a beam and passes round a travelling pulley, to the axle of which the weight, W, to be lifted is attached.

Let the applied force be F. Then the tension in the rope is F throughout.

$\therefore$ the pulley is sustained by two cords in each of which the tension is F.

$\therefore$ the weight supported is $2F$.

$$\therefore \quad W = 2F.$$

This takes no account of the weight of the pulley. If this be considerable and equal to w,

then $$W + w = 2F.$$

The above relation between F and W may be obtained by applying the principle of moments.

AB, the diameter of the pulley, may be regarded as a moving lever, since as the pulley rotates, one diameter is instantly replaced by another diameter.

If moments be taken about A, we have F acting at B and W at C.

$\therefore \qquad F \times AB = W \times AC.$

But $\qquad AB = 2AC.$

$\therefore \qquad W = 2F.$

27. Worked examples

Example 1. *An iron bar, AB, 80 cm long and weighing 3 kgf, turns about a fulcrum at A (Fig. 26). At B a weight*

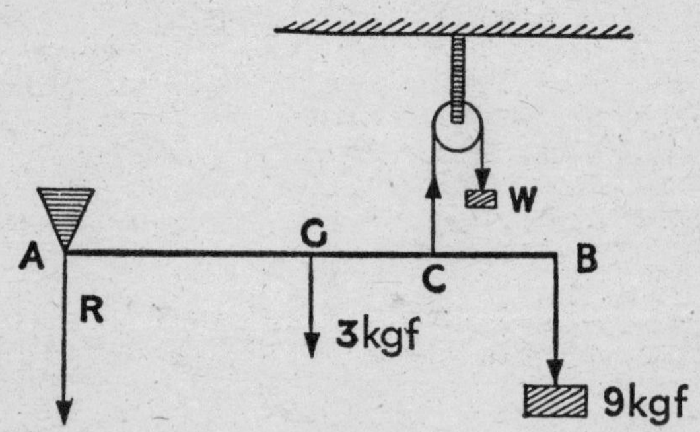

FIG. 26.

of 9 kgf is hung. At a point C, 20 cm from B, a cord is attached to the rod and is passed vertically upwards over a smooth pulley fixed to a beam. To this cord a load of W kgf is fixed to preserve equilibrium. Find W and the force on the rod at the fulcrum.

The tension in the cord passing over the pulley is the same throughout.

$\therefore \qquad$ at C, W kgf acts upward.

Let R = force **on the rod** at the fulcrum.

To find W take moments about A.

By doing this the unknown force R is eliminated. Then

	Anti-clockwise		**Clockwise**
	$W \times 60$	$=$	$(3 \times 40) + (9 \times 80)$.
$\therefore$	$60W$	$=$	$120 + 720$
$\therefore$	W	$=$	14 kgf.

To find R take moments about C.

Anti-clockwise **Clockwise**

$$(R \times 60) + (3 \times 20) = 9 \times 20.$$

$$\therefore \quad 60R + 60 = 180$$

$$R = 2 \text{ kgf.}$$

As a check:

Upward force $W = 14$ kgf.
Downward force $2 + 3 + 9 = 14$ kgf.

Example 2. *A uniformly loaded rectangular box of mass 360 kg is lying with a face on horizontal ground. A uniform crowbar, 120 cm long and weighing 12 kgf, is inserted 10 cm under it in a direction perpendicular to one edge and at the mid point of the edge. Find what force must be applied at the other end of the bar so that the box may be just tilted. Find also the force of the bar on the ground.*

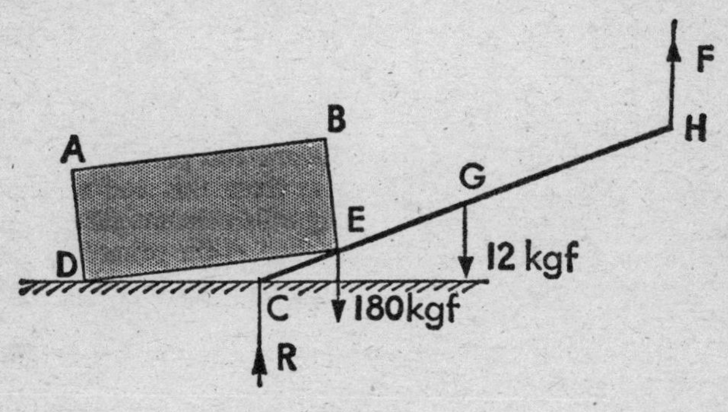

FIG. 27.

Fig. 27 represents a vertical section through the symmetrical centre of the box.

The bar, CH, when just on the point of tilting will sustain at E, the point of contact, 10 cm from C, a force which will be half the weight of the box—i.e. 180 kgf—the other half being borne by the other edge at D.

Let F be upward force applied at H.

Take moments about C for the equilibrium of the bar.

Anti-clockwise **Clockwise**

$$F \times 120 = (180 \times 10) + (12 \times 60)$$
$$= 2520 \text{ kgf cm}$$
$$\therefore \quad \mathbf{F = 21 \text{ kgf.}}$$

Let R be the force of the ground on the bar.
Take moments about H for the equilibrium of the bar.

Then $$R \times 120 = (180 \times 110) + (12 \times 60)$$
$$= 20\,520 \text{ kgf cm}$$
$$\therefore \quad \mathbf{R = 171 \text{ kgf.}}$$

Check:

Up $171 + 21 = 192$ kgf.
Down $180 + 12 = 192$ kgf.

Exercise 3

1. A heavy uniform bar, 1·5 m long and of 2 kg mass, is pivoted at one end. A weight of 30 kgf is attached at a point on the bar 0·15 m from this end. What applied force, 0·1 m from the other end, will just balance this weight?

2. A crowbar, 1·2 m long, has one end pivoted on the ground. At a distance of 0·2 m from this end a load exerts a force of 1000 kgf. Disregarding the weight of the crowbar, what force must be applied at the other end so as just to raise the load? What will be the thrust of the ground on the bar at its end?

3. A uniform bar 0·6 m long and weighing 2 kgf is pivoted at one end. A load of 6 kgf is applied at the other end. What upward force applied at a point 0·15 m from this will produce equilibrium? What will be the force at the fulcrum?

4. A man pulls an oar 2·1 m long with the end of the handle 0·6 m from the rowlock. Find the ratios of the pull at the end of the handle by the oarsman, the resistance of the boat to the oar at the rowlock, and the resistance of the water acting at the end of the oar.

5. A rectangular block of stone of mass 320 kg lies horizontally on the ground. A crowbar, of length 2 m

and weight 15 kgf is pushed a distance of 0·4 m under the block. What force applied at the other end of the crow-bar will just tilt the block?

6. In a pair of nutcrackers a nut is placed 15 mm from the hinge and the pressure applied at the handles is estimated to act at a distance of 120 mm from the hinge, and to be equal to 3·5 kgf. What force is applied to the nut?

7. A heavy uniform plank, 8 m long and weighing 40 kgf, projects 2 m horizontally from the top of the cliff. How far can a man of mass 80 kg move along the plank before it tips up?

8. In a safety valve (*see* Fig. 21(*b*)) the distance between the fulcrum and the centre of the piston is 60 mm; the area of the surface of the valve is 1200 mm^2. The lever is 160 mm long and has a load of 60 kgf at the end of it. Find the pressure of the steam per mm^2 when the valve is just beginning to move upwards.

9. A pump handle is 1·2 m long from the pivot to the end. The pivot is 0·1 m from the point where it is attached to the plunging rod. A force of 10 kgf is applied at the end of the handle. What force is applied to the plunging rod?

10. A uniform bar is 8 m long and is 10 kg in mass. It is pivoted at one end and a load of 40 kgf is applied 1 m from the pivot. What force applied at the other end of the bar will just support this load?

CHAPTER III

CENTRE OF GRAVITY

28. Centre of parallel forces

In the previous chapter it was shown that, if two parallel forces represented by weights acted on a bar so that their opposing moments about an axis on the bar were equal, the bar was in equilibrium.

It was also seen that the two forces could be replaced by a single force, equal to their sum and parallel to them, acting at the axis. This force is called the **resultant of** the forces, and the point at which it acts is called the **centre of force.**

This principle may be extended to any number of forces.

Suppose that F_1, F_2, F_3 are parallel forces.

Let R_1 be the resultant of F_1 and F_2.

Then $$R_1 = F_1 + F_2.$$

R_1 acts at a point which divides the distance between F_1 and F_2 in the ratio $F_2 : F_1$.

Now, R_1 and F_3 being parallel forces, the resultant of these can be found. Let R be the resultant.

Then $$R = R_1 + F_3 = F_1 + F_2 + F_3.$$

It acts at a point between R_1 and F_3 which divides the distance between them in the ratio of $F_3 : (F_1 + F_2)$.

This is the point where R, the **resultant of F_1, F_2, and F_3, acts, and is the centre of force of the system.**

This may be extended to any number of forces.

When the forces acting are due to gravity, the centre of them is called the centre of gravity of the system.

Now, any solid body can be considered as composed of a large number of particles upon each of which the force of gravity acts. All these forces are parallel, and their resultant, expressed by the **weight of the body,** acts at the centre of these forces, which is therefore the **centre of gravity of the body,** thus:

The centre of gravity of a body is the point through which the resultant of the earth's pull upon the body passes and at which the weight of the body can be considered as acting.

29. To find the centre of gravity of a number of particles

Let A and B (Fig. 28) be two particles of weights W_1 and W_2.

By the principle of § 24 the C.G. of these is at G_1 on the straight line joining them where

$$\frac{AG_1}{BG_1} = \frac{W_2}{W_1}.$$

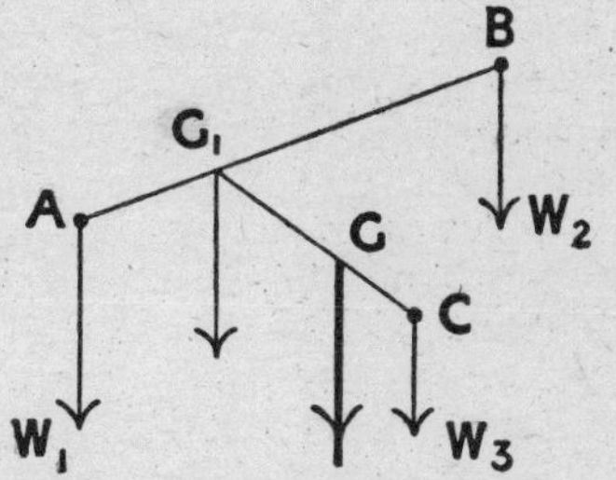

$\therefore$ as we have seen, we can regard $W_1 + W_2$ as acting at G_1.

Let C be a third particle of weight W_3.

Then the C.G. of $W_1 + W_2$ at G_1 and W_3 at C lies at G on the straight line joining G_1C.

FIG. 28.

Where $$\frac{CG}{G_1C} = \frac{W_1 + W_2}{W_3}.$$

The point G is therefore the centre of gravity of the three particles, and their resultant $W_1 + W_2 + W_3$ can be considered as acting at it.

30. Centre of gravity of a uniform rod

We have assumed that a uniform rod can be balanced about its centre point, and experience shows that it can be supported in equilibrium by resting it on a fulcrum at this point. This implies that the effect of the force of gravity upon the parts of the rod on the two sides of the fulcrum is the same.

Let DE (Fig. 29) represent a uniform rod, and let A and

B be two small equal portions of the rod situated at equal distances from the fulcrum, *C*, at the centre of the rod. Moments of these pieces about the fulcrum must be equal.

∴ they could be represented by a single force equal to their sum, acting at *C*.

The same reasoning applies to all such pieces throughout the equal arms.

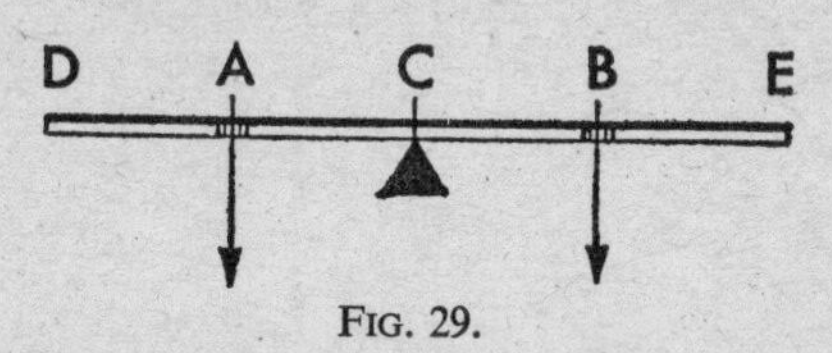

FIG. 29.

∴ all the weights of all such parts in these two arms can be replaced by a single weight, equal to their sum—i.e. the weight of the rod acting at *C*.

Hence *C* **is the centre of gravity of the rod.**

A similar result holds for a narrow rectangular strip.

31. Centre of gravity of regular geometrical figures

Figures which are geometrically symmetrical, such as a rectangle, circle, equilateral triangle, have what may be called a symmetrical centre, as, for example, the intersection of the diagonals of a rectangle. Such a point will be the centre of gravity of the figures, if uniform, since the forces of gravity will balance about it.

32. Experimental determination of the centre of gravity

If a body be suspended from a point near one of its boundaries it will remain at rest when the point of support of the string lies vertically above the centre of gravity. In this way only can the principle of balance of turning moments be satisfied. This enables us to obtain experimentally the centre of gravity in certain cases.

The easiest object to experiment with is a flat uniform piece of cardboard, of any shape, such as is suggested by Fig. 30.

Pierce it near a rim (*A* in Fig. 30) with a fine needle, and stick this on a vertical board or paper. From this needle suspend a fine thread with a small weight attached and let it hang vertically. Mark two points on the straight line formed where the line touches the board, and draw the straight line *AX*.

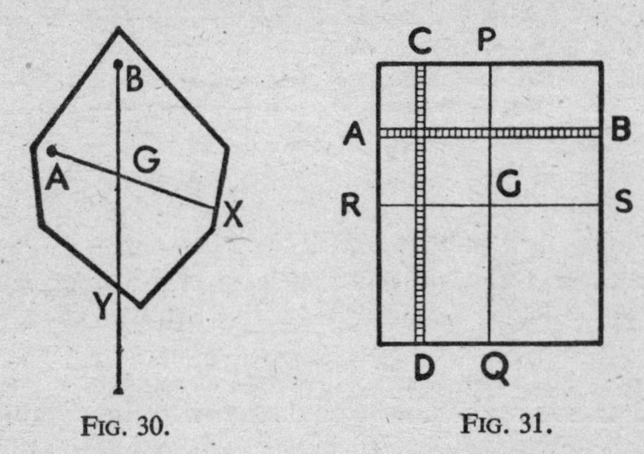

FIG. 30. FIG. 31.

Then the centre of gravity must lie on *AX*.

Take a second point *B*, repeat the experiment, and obtain the straight line *BY*. The centre of gravity must also lie on *BY*.

∴ the **centre of gravity lies at *G*,** the intersection of the two straight lines.

If the cardboard or other lamina has a perceptible thickness and is uniform, the centre of gravity will lie half-way between the two surfaces, underneath the point *G*.

For obvious reasons, this method of finding the C.G. cannot be employed with solid bodies, but it should be remembered that, when *any* body is suspended at a point and remains permanently at rest, the C.G. lies vertically below this point.

33. Centre of gravity of a rectangular lamina

The C.G. of a rectangle will be the intersection of the diagonals (§ 31), since it is a symmetrical geometrical figure. But we employ this shaped lamina to demonstrate a very useful method of finding the C.G.

The rectangle can be considered as being composed of a large number of narrow rectangles such as *AB* (Fig. 31), all parallel to a pair of opposite sides. The C.G. of *AB* is at the centre of the strip as shown in § 29. The C.Gs. of all such strips must therefore lie along the straight line *PQ*, which joins the middle points of all the strips, and *P* and *Q*, the mid points of the opposite sides.

Similarly, the rectangle can be considered as made up of strips, such as *CD*, parallel to the other pair of sides. The C.G. of all such strips will lie along *RS*, which joins the mid points.

∴ **the C.G. lies at *G*, the intersection of *PQ* and *RS*.**

This point *G* is also the intersection of the diagonals.

Similarly we may find the C.Gs. of a square, a parallelogram and a rhombus.

34. Centre of gravity of a triangular lamina

The method is similar to that employed for a rectangle. The triangle *ABC* (Fig. 32) is regarded as composed of a large number of very narrow strips, such as *DE*.

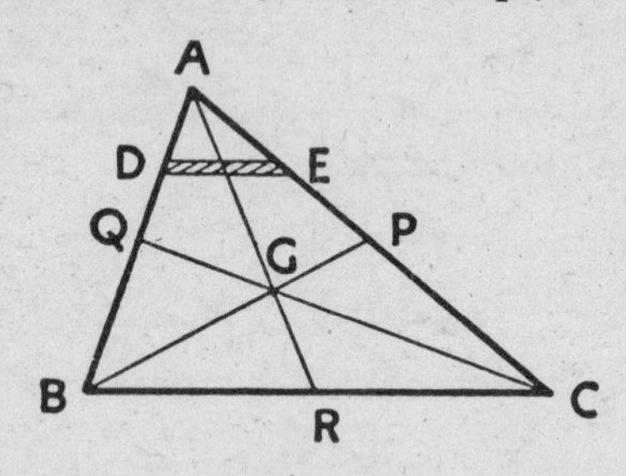

FIG. 32.

This strip, being very narrow, may be regarded as having its C.G. at the centre of the strip.

The centres of all such strips parallel to it will therefore lie on the straight line joining *A* to *R*, the centre of *BC*. This straight line *AR* is a median of the triangle.

Similarly the centres of all such strips parallel to *AC* lie on the median *BP*.

∴ the C.G. of the triangle lies at G, the intersection of the medians.

It follows that the third median, CQ, must pass through G and contains the C.Gs. of all strips parallel to AB.

In Geometry it is proved that G is one of the points of trisection of each median.

$$\therefore \quad BG = 2PG$$
$$AG = 2RG$$
$$CG = 2QG.$$

35. Centre of gravity of equal particles at the angles of a triangle

It is useful to note that the C.G. of a triangular lamina is the same as that of three particles, each of which has a weight which is one-third of that of the triangle, placed at the angular points of the triangle.

Let the weight of ΔABC (Fig. 33) be W.

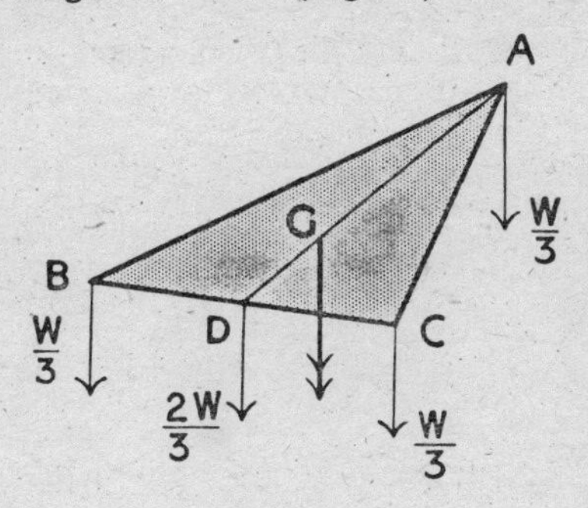

FIG. 33.

Let particles each of weight $\frac{W}{3}$ be placed at the angular points.

As we have seen (§ 29), particles of $\frac{W}{3}$ acting at B and C are equivalent to $\frac{2W}{3}$ acting at D, where D is the mid point of BC.

$\therefore$ the C.G. of the system must be same as that of $\frac{2W}{3}$ acting at D and $\frac{W}{3}$ acting at A.

But C.G. of $\frac{2W}{3}$ at D and $\frac{W}{3}$ at A is

$$\frac{2W}{3}+\frac{W}{3} \text{ at } G,$$

where AD is a median and

$$AG:GD = \frac{2W}{3}:\frac{W}{3} = 2:1.$$

Thus G must be the C.G. of the three particles. But it is also the C.G. of the ΔABC.

$\therefore$ the two C.Gs. coincide.

36. Centre of gravity of composite bodies

It is often required to find the centre of gravity of a lamina composed of two or more regular figures. If the centre of gravity of each of these is known, the centre of gravity of the whole figure can be found by methods shown in the following examples:

Example 1. *To find the centre of gravity of a uniform lamina consisting of a square and an equilateral triangle constructed on one side.*

In problems connected with lamina of uniform structure and thickness, the mass, and consequently the weight, of a lamina is proportional to its surface area.

$\therefore$ when taking moments the areas may be used as **representing** the actual weights.

In Fig. 34 G_1, the intersection of diagonals, is the C.G. of the square.

Also EG_1 is an axis of geometrical symmetry for the whole figure.

$\therefore$ C.G. of the Δ and of the composite figure will lie on this line.

But C.G. of Δ is at G_2, where FG_2 is $\frac{1}{3}$ of the median EF.

If $\qquad a =$ length of side of square,

$\therefore \qquad a^2 =$ area of square;

also $$EF = a \times \frac{\sqrt{3}}{2}$$

$$FG_2 = \frac{1}{3} a \times \frac{\sqrt{3}}{2} = \frac{a\sqrt{3}}{6}$$

and $$\Delta AEB = \frac{1}{2} a \times \frac{a\sqrt{3}}{2} = \frac{a^2\sqrt{3}}{4}.$$

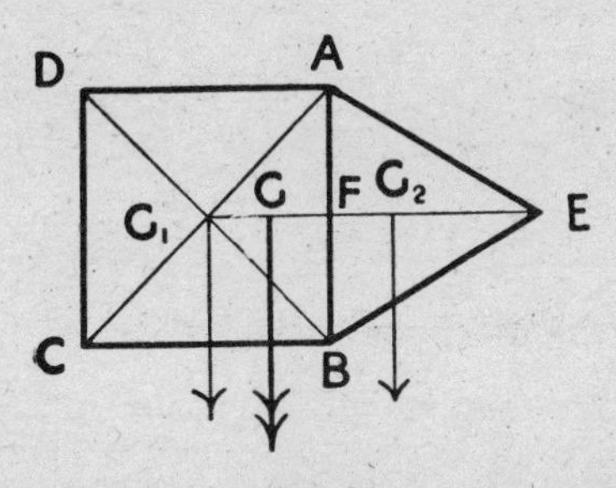

FIG. 34.

Taking the weight of each figure as concentrated at its C.G. we have:

(1) a^2 acting at G_1.

(2) $\frac{a^2\sqrt{3}}{4}$ acting at G_2.

$\therefore$ C.G. of the whole is at G, where

G_1G_2 is divided at G in the ratio $a^2 : \frac{a^2\sqrt{3}}{4}$,

i.e. in the ratio $4 : \sqrt{3}$.

Example 2. *To find the centre of gravity of a quadrilateral lamina.*

First method. The following method is useful for finding the C.G. by drawing.

(1) In the quadrilateral $ABCD$ (Fig. 35) draw the diagonal AC.

Consider Δs ABC, ADC.

Let weights be W_1 and W_2.

Take E, the mid point of AC. Join DE, BE.

C.G. of ADC is at G_1, where $EG_1 = \frac{1}{2}ED$.

C.G. of ABC is at G_2 where $EG_2, = \frac{1}{3}EB$.

Join G_1G_2.

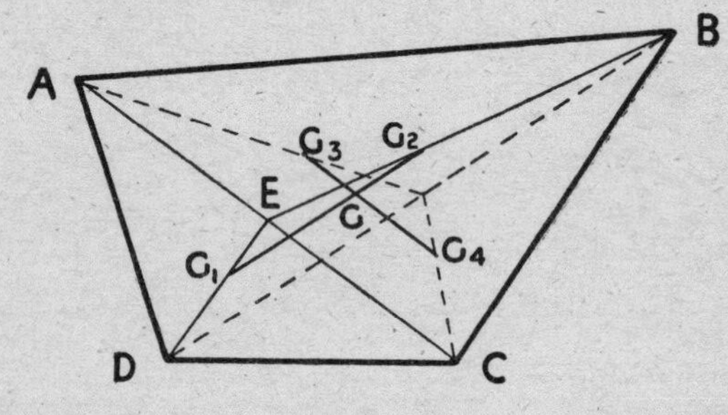

FIG. 35.

Then C.G. of the whole quadrilateral is at G, where $G_1G:G_2G = W_1:W_2$.

(2) Draw the diagonal BD (dotted line) and so divide the quadrilateral into Δs ABD, DBC.

Now proceed as before to join the C.G.s these Δs, viz. G_3 and G_4 (dotted lines used throughout).

$\therefore$ C.G. of quadrilateral lies on G_3G_4, but it also lies on G_1G_2.

$\therefore$ it lies at G, their point of intersection.

Second method. This method is similar to that in § 35 for a triangle.

In Fig. 36, let weight of $\Delta ABD = W_1$.

In Fig. 36, let weight of $\Delta BCD = W_2$.

C.G. of ΔABD is same as $\frac{W_1}{3}$ at points A, B, D.

C.G. of ΔBCD is same as $\frac{W_2}{3}$ at points B, C, D.

$\therefore$ **C.G. of quadrilateral is the same as:**

$$\frac{W_1}{3} \text{ at } A,$$

$$\frac{W_1}{3} + \frac{W_2}{3} \text{ at } B,$$

$$\frac{W_2}{3} \text{ at } C,$$

$$\frac{W_1}{3} + \frac{W_2}{3} \text{ at } D.$$

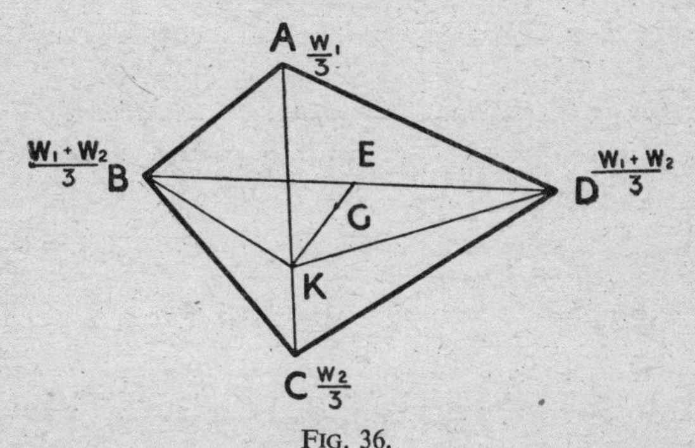

FIG. 36.

Let AC be divided at K in ratio $W_1 : W_2$—i.e. wts. of Δs. Then $\frac{W_1}{3}$ at A and $\frac{W_2}{3}$ at C are equivalent to $\frac{W_1}{3} + \frac{W_2}{3}$ acting at K.

Since we have $\frac{W_1 + W_2}{3}$ acting at points B, K, D,

$\therefore$ **C.G. of quadrilateral is the same as that of ΔBKD.**

$\therefore$ if E is mid point of BD,

C.G. of quadrilaterial is at G on EK, where $EG = \frac{1}{3}EK$.

37. Use of moments in finding centre of gravity

One of the most valuable and effective methods of finding the centre of gravity is by an application of the principle of moments.

If a body or system of bodies is made up of a number of parts whose weights and centres of gravity are known, the **sum of these parts is their resultant, and it acts at the centre of gravity of the whole.**

Consequently **the moment about any axis of the resultant acting at the centre of gravity of the whole is equal to the sum of the moments of the parts acting at their centres of gravity.**

This can be expressed generally as follows:

Let $w_1, w_2, w_3 \ldots$ be a series of weights whose distances from a given axis are $d_1, d_2, d_3 \ldots$

The **sum of their moments** about this axis is

$$w_1d_1 + w_2d_2 + w_3d_3 + \ldots$$

Let $\bar{x}$ be distance of C.G. of the whole from the axis about which moments are taken.

Then **moment of the resultant** about the axis is

$$(w_1 + w_2 + w_3 + \ldots) \times \bar{x}.$$

But moment of resultant = moments of the parts.

$\therefore (w_1 + w_2 + w_3 + \ldots) \times \bar{x} = w_1d_1 + w_2d_2 + w_3d_3 + \ldots$

$$\therefore \qquad \bar{x} = \frac{w_1d_1 + w_2d_2 + w_3d_3 + \ldots}{w_1 + w_2 + w_3 + \ldots}$$

i.e. **dist. of C.G. from axis** $= \dfrac{\textbf{sum of moments of parts}}{\textbf{sum of weights of parts}}$.

Similarly, if any other axis were taken, and if $\bar{y}$ is the distance of the C.G. from the axis, $\bar{y}$ can be determined in the same way.

38. Worked examples

Example 1. *A uniform rod AB is* 60 *cm long and weighs* 3 *kgf. Weights are placed on it as follows:* 2 *kgf at* 20 *cm from A,* 5 *kgf at* 40 *cm from A, and* 6 *kgf at B. How far is the centre of gravity of the whole system from A?*

The arrangement of the forces is shown in Fig. 37.

Take moments about an axis at *A*. **The sum of moments of the three weights and the weight of the rod** acting at the centre of the rod must equal the **moment**

of the resultant acting at the unknown centre of gravity of the whole.

Let $\bar{x}$ = distance of the C.G. from A.
Resultant = sum of weights
$= 2 + 5 + 6 + 3$
$= 16$ kgf.

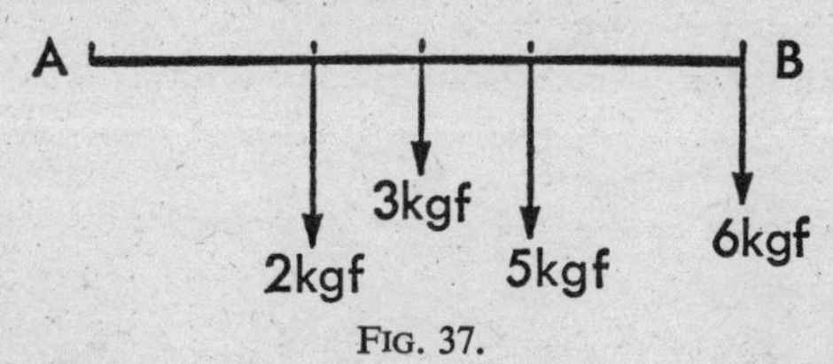

FIG. 37.

Equating moments about A:

$$16 \times \bar{x} = (2 \times 20) + (5 \times 40) + (6 \times 60) + (3 \times 30)$$
$$= 690$$
$$\therefore \quad \bar{x} = \frac{690}{16} = 43 \text{ cm from } A.$$

Example 2. *Find the centre of gravity of a thin uniform lamina as shown in Fig.* 38.

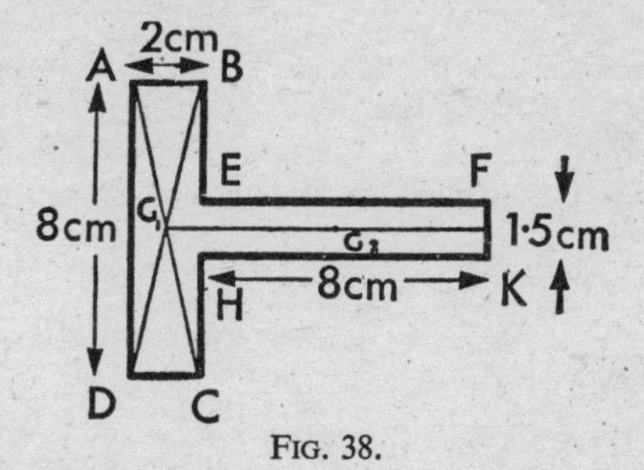

FIG. 38.

Let the weight of a sq cm of the lamina be w gf.
Wt. of $ABCD = (8 \times 2)w = 16w$ gf.
C.G. of $ABCD$ is at G_1, where G_1 is the intersection of the diagonals and $OG_1 = 1$ cm, where O is the centre of AD.
Wt. of $EFKH = (8 \times 1\frac{1}{2})w = 12w$ gf.

C.G. of *EFKH* is at G_2, where $OG_2 = 6$ cm and G_1G_2 is the symmetrical axis of the figure.

$\therefore$ Total weight $= 16w + 12w = 28w$.

This acts at an unknown C.G.

Let distance of C.G. from O be $\bar{x}$.

Taking moments about *AD*.

$$28w \times \bar{x} = (16w \times 1) + (12w \times 6)$$
$$= 88w$$
$$\bar{x} = \frac{88w}{28w} = \frac{22}{7} \text{ cm.}$$

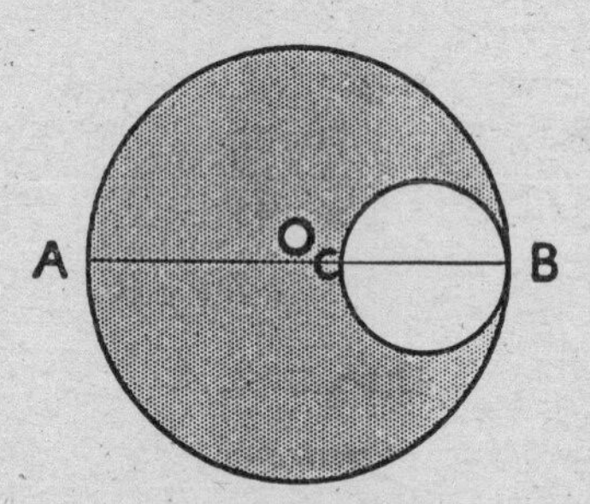

FIG. 39.

Example 3. *Find the centre of gravity of a thin uniform circular metal plate of radius* 12 *cm, when there has been cut out a circular piece of metal of radius* 4 *cm which touches the circumference of the plate.*

Let *AB* (Fig. 39) be a diameter of circle.

Let *CB* be a diameter of circle cut out.

By symmetry C.G. of remainder lies on *AB*.

Let w gf = weight of plate per sq cm.

Let G be position of the required centre of gravity.

Weight of whole circle $= (\pi \times 12^2 \times w)$ gf.

Weight of cut-out circle $= (\pi \times 4^2 \times w)$ gf.

Weight of remainder $= \pi(144 - 16)w$ gf.

$= 128\pi w$ gf.

Now the moment of the weight of the **whole** circle about any axis must equal the sum of the moments of the circle cut out and the remainder.

I.e. moment of whole circle at O = moment of remainder at G + moment of circle cut out acting at the centre.

$\therefore$ moment of remainder at G = moment of whole circle at O − moment of circle cut out.

Taking moments about A.

$$128\pi w \times AG = 144\pi w \times 12 - 16\pi w \times 20.$$

$$\therefore \quad 128AG = 144 \times 12 - 16 \times 20 = 1408.$$

$$\therefore \quad AG = \frac{1408}{128} = 11 \text{ cm.}$$

Example 4. *Weights of 2 kgf, 4 kgf, 6 kgf, 4 kgf are placed at the corners A, B, C, D of a uniform square lamina of weight 4 kgf. The side of the square is 120 cm. Find the distance of the centre of gravity of the whole from AB and AD.*

Fig. 40 represents the square and the arrangement of the weights.

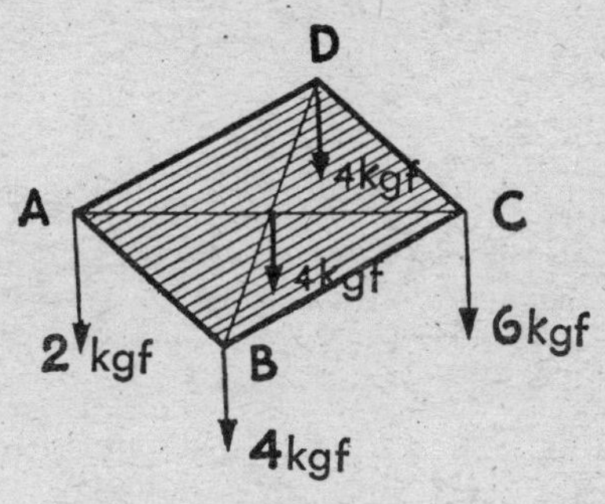

FIG. 40.

The C.G. of the lamina itself is at O, the intersection of the diagonals.

(1) **Taking moments about AB.**

Let $\bar{x}$ = distance of C.G. of the whole from AB.

The turning moments about the axis AB of weights at A and B, on the axis itself, will be zero.

Weight of the whole system = $2 + 4 + 6 + 4 + 4$ = **20 kgf.**

Now, moment of the resultant—i.e. from 20 kgf acting at unknown C.G. = sum of moments of parts of the system.

$$\therefore \quad 20 \times \bar{x} = (6 \times 120) + (4 \times 120) + (4 \times 60)$$
$$= 1440$$
$$\therefore \quad \bar{x} = \frac{1440}{20} = 72 \text{ cm.}$$

(2) **Taking moments about *AD*.**

Let $\bar{y}$ = distance of C.G. of whole from AD.
Then the equation of moments, found as before, is:

$$20 \times \bar{y} = (4 \times 60) + (6 \times 120) + (4 \times 120)$$
$$= 1440$$
$$\therefore \quad y = 72 \text{ cm.}$$

∴ The **centre of gravity is 72 cm from *AB* and *AD*.**

Check: distance of C.G. from *CB* and *CD* could be similarly calculated.

Exercise 4

1. On a light rod, *AB*, 3 m long, weights of 3 kgf and 5 kgf are hung at distances of 1 m and 2·5 m from *A*. Where will the centre of gravity be?

2. If the rod in the last question were uniform and of weight 2 kgf, where would the centre of gravity of the whole then be?

3. On a light rod, *AB*, 1·25 m long, weights of 2 kgf, 4 kgf, 6 kgf and 8 kgf are hung at intervals of 0·25 m between *A* and *B*. Find the centre of gravity of the whole.

4. On a light rod, *AB*, 1·25 m long, weights of 2 kgf, 6 kgf and 4 kgf are placed at distances of 0·1 m, 0·2 m and 0·5 m respectively from *A*. What weight must be placed 0·8 m from *A* if the centre of gravity is to be 0·49 m from *A*?

5. *ABC* is a light isosceles triangle. $AC = AB = 3$ m, and $BC = 2$ m. Weights of 1 kgf, 2 kgf, 2 kgf are hung at *A*, *B* and *C* respectively. How far is the centre of gravity of these from *BC*?

6. *ABC* is a triangle of weight 2 kgf. Weights of 2 kgf, 4 kgf, and 4 kgf are placed at *A*, *B* and *C* respectively. What is the position of the centre of gravity of the whole?

7. Find the centre of gravity of a uniform piece of cardboard consisting of a square *ABCD* of side 0·6 m, and an isosceles triangle *BCE* constructed on side *BC* of the square and having an altitude *FE* 0·4 m long.

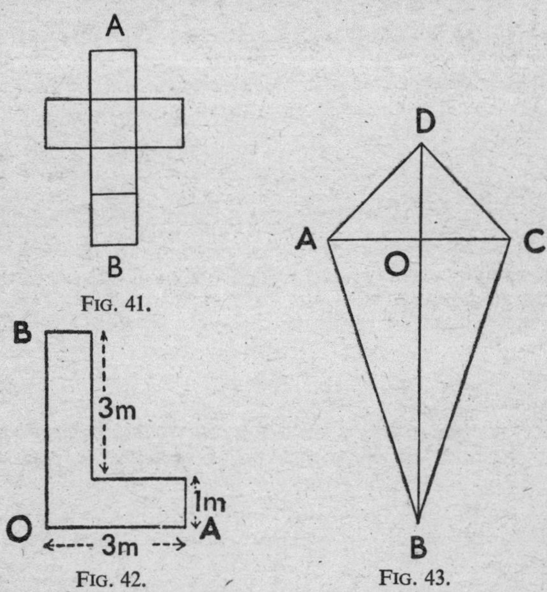

FIG. 41.

FIG. 42.

FIG. 43.

8. Fig. 41 represents a cross containing six squares, each 2 m square. It is made of thin uniform metal. Find the centre of gravity of the cross in relation to *AB*.

9. Find the centre of gravity of a uniform lamina of the shape and dimensions shown in Fig. 42, giving its distance from *OA* and *OB*.

10. The lamina shown in Fig. 43 consists of two isosceles triangles *ABC*, *ADC*. The diagonals *AC* and *BD* are 0·6 m and 0·8 m respectively, and *OD* is 0·2 m. Find the distance of the centre of gravity of the whole from *B*.

11. Find the centre of gravity of six thin uniform metal discs, arranged as shown in Fig. 44. The diameter of each circle is 10 mm.

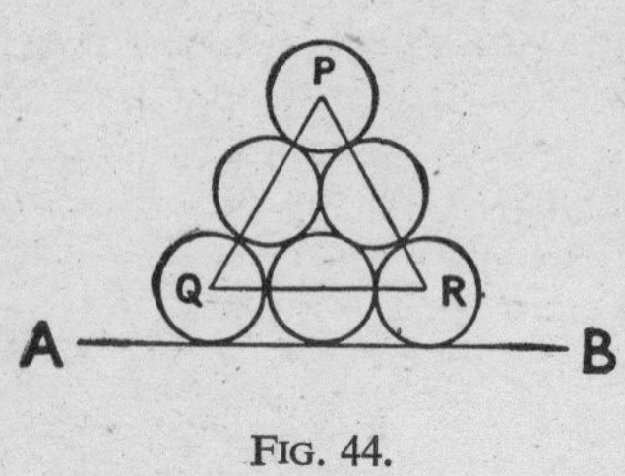

FIG. 44.

12. Weights of 2 kgf, 3 kgf, 4 kgf and 5 kgf are placed at the corners *A*, *B*, *C*, *D* respectively, of a square of side 0·1 m and weighing 2 kgf. Find the position of the centre of gravity in relation to *AB* and *AD*.

39. Centre of gravity of regular solids

The centres of gravity of a few regular solids are given below, in most cases without proof, as these require more advanced mathematics to be satisfactory.

Symmetrical solids. The centres of gravity of those solids which are symmetrical bodies, such as cylinders, spheres, etc., are at the geometrically symmetrical centres, since the mass of a symmetrical solid will balance about that point.

Rectangular prism. This is a symmetrical body, but a proof, similar to that employed for a lamina, is given because of the general usefulness of the method.

Fig. 45 shows a rectangular right prism in which is indicated a very thin lamina, cut at right angles to the axis of the prism, and therefore parallel to the two bases. This lamina is a rectangle in shape, and its centre of gravity is at the centre of it—that is, at the intersection of the diagonals. It will therefore lie on the axis of the prism, which is the straight line joining the intersections of the diameters of the bases. If the prism is conceived as being made up of a large number of such laminæ, the centre of gravity of them all will lie on the axis. From the symmetry of the solid the centre of gravity will therefore be at the **middle point of the axis.**

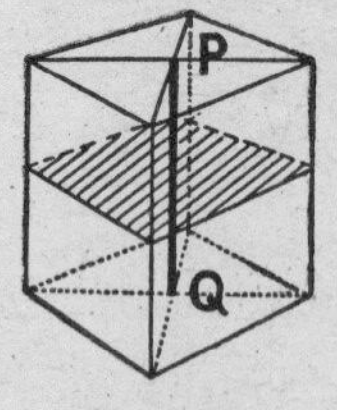

FIG. 45.

The **cylinder** is a special case of the prism, a section, at right angles to the axis, always being a circle. The centre of gravity is therefore at the centre of the axis.

Sphere. The centre of gravity of a sphere is obviously at the centre.

Hemisphere. The centre of gravity of a hemisphere requires more advanced mathematics for its determination than is assumed in this book. It lies on the radius drawn perpendicular to the base from the centre and is three-eighths of the length of the radius from the base. If r be the radius, the

C.G. is $\frac{3}{8}r$ from centre of base.

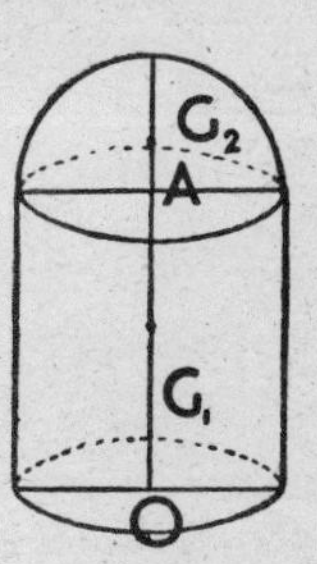

FIG. 46.

Right pyramids and cone. In all of these the centre of gravity lies at a point one-fourth the way up the axis from the centre of the base.

If h be the length of this axis, that is, the **altitude** of the pyramid or cone,

C.G. is $\frac{h}{4}$ from the centre of the base.

40. Worked example. *A cylinder of height* 10 *cm and radius of base* 4 *cm is surmounted by a hemisphere of the same radius as that of the base of the cylinder, and made of the same material. Find the distance of the centre of gravity of the composite body from the centre of the base of the cylinder.*

In Fig. 46, G_1, the mid point of OA, the axis of the cylinder is the C.G. of that body. G_2 is the C.G. of the hemisphere and $AG_2 = \frac{3}{8}r$, where r is the radius.

$$\begin{aligned}
\text{Vol. of cylinder} &= \pi r^2 h \ (h = \text{the height}) \\
&= \pi \times 4^2 \times 10 = 160\pi \text{ cm}^3. \\
\text{Vol. of hemisphere} &= \tfrac{2}{3}\pi r^3 \\
&= \tfrac{2}{3}\pi \times 64 = \frac{128\pi}{3} \text{ cm}^3. \\
\therefore \text{Total vol.} &= 160\pi + \frac{128\pi}{3} = \frac{608\pi}{3} \text{ cm}^3.
\end{aligned}$$

Also $\quad OG_1 = \frac{h}{2} = 5$ cm

$\quad OG_2 = 10 + (\frac{3}{8} \times 4) = \frac{23}{2}$ cm.

Let $\bar{y}$ = distance of C.G. of whole from O.

Since the bodies are of the same material, the weights are proportional to their volumes.

Taking moments about *O*.

$$\frac{608\pi}{3} \times \bar{y} = (160\pi \times 5) + \left(\frac{128\pi}{3} \times \frac{23}{2}\right)$$

$$\therefore \quad \bar{y} \times \frac{608}{3} = 800 + \frac{1472}{3} = \frac{3872}{3}$$

$$\therefore \quad \bar{y} = \frac{3872}{3} \times \frac{3}{608}$$

$$= \mathbf{6{\cdot}37\ cm} \text{ approximately.}$$

41. Equilibrium of bodies

We have seen (§ 32) that a body which is suspended from a point on or near a boundary will be in equilibrium when the point of suspension and the centre of gravity are in the same vertical line (Fig. 47(*a*)). Similarly, if a

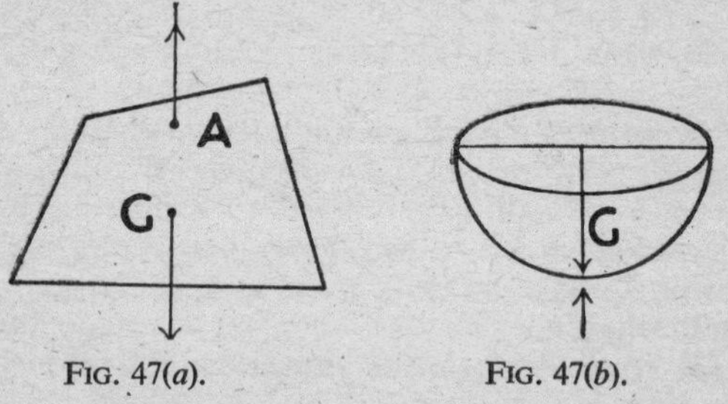

FIG. 47(*a*). FIG. 47(*b*).

body is resting on a surface (Fig. 47(*b*)), the centre of gravity and the point of support are in the same straight line. In each case the body is acted upon by:

(1) Its weight acting vertically downwards at the centre of gravity.

(2) An equal and opposite force acting vertically upward at the point of suspension or point of support.

42. Stable, unstable, and neutral equilibrium

Suppose that a body which is suspended in equilibrium receives a slight displacement (Fig. 48(*b*)). Three cases must be considered:

(1) **When the point of suspension is above the C.G.** Then the weight acting at the centre of gravity *G*, Fig. 48(*a*), being no longer in a vertical line with the force *T* at the point of suspension, *A*, exerts a turning moment, $T \times AB$, which tends **to swing the body back to its original position.** In such a case the body is said to be in **stable equilibrium.**

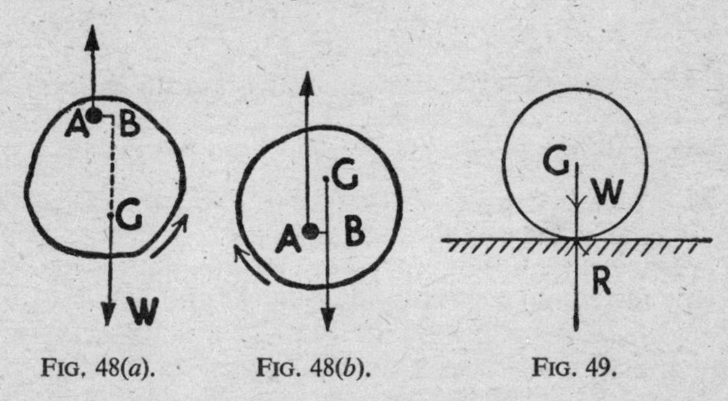

FIG. 48(*a*). FIG. 48(*b*). FIG. 49.

(2) **When the point of support is below the centre of gravity.** In this case, as will be seen from Fig. 48(*b*), the weight acting at *G* exerts a turning moment which tends to turn the body away from its original position. The body is then said to be in **unstable equilibrium.**

(3) **When the point of suspension is at the centre of gravity,** as, for example, of a circular lamina at its centre, or a wheel at its axis. Any displacement of the body in its own vertical plane will result in it remaining at rest in its new position.

Such a body is said to be in **neutral equilibrium.**

If the body is resting on a horizontal surface, such as a billiard ball on a table (Fig. 49), the forces acting are the weight of the ball downwards and the thrust

(R) of the table upwards. If there is a displacement, the ball will tend to come to rest with its centre of gravity, G, vertically over a new point of support.

43. Definitions of equilibrium and examples

(1) Stable equilibrium

A body is said to be in stable equilibrium when, on receiving a slight displacement, it tends to return to its original position.

Any body with a relatively large base, resting on a horizontal support and with a comparatively low centre of gravity, is stable. The following are a few examples: a right prism or a cylinder resting on a base; a cone resting on its base; this book resting on its side.

A necessary condition of stability is that the centre of gravity is vertically above what may be described as the contour of its base. In the case of a body with more than one support, this includes not only the area of the ground covered by each support, but also the area between them. For example, a table with four legs is stable because of the large area included between the legs and the straight lines joining them. This is important in many of the actions of life which depend on accurate balancing. In the case of a man standing up, the contour of his base is not only the actual area of the surface covered by his feet, but also that area between them and straight lines drawn from toe to toe and heel to heel. So long as the man's centre of gravity is vertically over this area he maintains his balance easily. That is why in cases where there is lateral swaying, as in a bus, a train, or a ship, a man instinctively places his feet farther apart, and so increases his base contour. The student may easily work out for himself problems of balancing connected with riding a bicycle, walking on a narrow ledge, etc.

(2) Unstable equilibrium

A body is said to be in unstable equilibrium when, on receiving a slight displacement, it tends to go farther away from its position of rest.

A body with a small base and a high C.G. is usually

unstable. Examples are: a cone resting on its vertex from which a small piece has been cut off, a lead pencil balanced on its base, a narrow book standing upright on a table, etc.

(3) Neutral equilibrium

A body is in neutral equilibrium when, on receiving a slight displacement, it tends to come to rest in its new position.

Examples are a ball, a cylinder lying on its curved surface, a hemisphere lying on its curved surface, a cone lying on its oblique surface, etc.

Exercise 5

1. A cone of altitude 60 mm is placed on top of a cylinder whose base is the same area as the base of the cone and whose height is also 60 mm. The axes of the bodies are in the same straight line. How far is the centre of gravity of the composite figure from the base of the cylinder? Both bodies are of the same material.

2. A cone of height 0·4 m and of base radius 0·2 m is fastened to the flat surface of a hemisphere whose base has the same area as the base of the cone. If the two bodies are made of the same material, find the centre of gravity of the composite body.

3. A uniform piece of wire is bent to the shape of an isosceles triangle, with a base of 160 mm and each of the equal sides being 100 mm. Find the centre of gravity of the triangular wire.

4. From a uniform circular disc of diameter 1·2 m and mass 4 kg, a circular piece 0·3 m in diameter is removed, the shortest distance between the circumferences of the circles being 0·1 m. Find the position of the centre of gravity of the remainder.

Hint. Remember that the masses of the discs are proportional to the squares of their diameters.

5. From a rectangular lamina 0·1 m by 0·08 m a square of 0·04 m side is cut out of one corner. Find the distance

of the centre of gravity of the remainder from the uncut sides.

6. From one of the corners of an equilateral triangle of side 40 mm an equilateral triangle of 20 mm side is removed. Determine the centre of gravity of the remaining figure.

7. State in the following cases whether the equilibrium is stable, unstable or neutral:

(1) A door which swings about hinges with a vertical axis.

(2) A book lying on its side on a table with its centre of gravity just vertically over the edge of the table.

(3) A spherical marble lying in a basin with a spherical-shaped bottom.

(4) A ladder 9 m long resting against a wall, with its foot on the ground, and 0·1 m from the wall.

CHAPTER IV

RESULTANT OF NON-PARALLEL FORCES
THE PARALLELOGRAM OF FORCES

44. Geometric representation of a force

When investigating theorems and problems connected with forces we find it convenient to represent a force completely by a straight line.

For this purpose the straight line must show:

(1) The **direction** of its line of action, in relation to some fixed direction.

(2) The **sense** in which it acts along the line of action.

(3) The **magnitude**, shown by the length of the line, measured on a suitable scale.

Thus in Fig. 50, PQ represents a force of 6 units acting at a point O, making an angle of 25° with the direction of $X'OX$. The **sense** in which it acts along

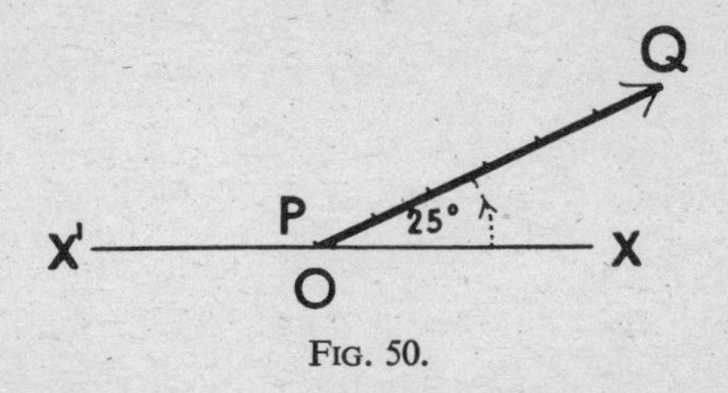

Fig. 50.

PQ is **from** P **to** Q; this is indicated by the arrow-head, and also by the order of the letters when describing it as the force PQ. If it acted from Q to P we should describe it as the force QP, and the arrow-head would be reversed.

45. Vector quantity

A quantity, such as a force, which possesses both **direction and magnitude** is called a **vector quantity,** and the straight line by which it is represented, as *PQ* above, is called a **vector.**

Other examples of vector quantities with which we shall be concerned in later chapters are displacements, velocities and accelerations.

Vectors are employed in the graphical solution of many problems, and the student would do well to acquire some elementary knowledge of them. This can be found in books on Practical Mathematics, such as *National Certificate Mathematics* which is published by the English Universities Press.

46. Non-parallel forces acting on a body

The forces with which we were concerned in the previous chapters have been parallel and acting in the same direction. We have seen how a number of such forces acting on a body can be replaced by a single force called the **resultant,** whose magnitude is equal to the algebraic sum of the magnitudes of the separate forces.

We now proceed to consider forces which are not parallel but concurrent; we must ascertain if such forces can have a resultant and how it may be obtained.

A practical example of the problem involved may help in understanding our object.

Suppose that two men were pulling down a tree by means of ropes attached to it at the same height. If they were to pull on ropes which were parallel, or if they pulled in the same direction on the one rope, the total force exerted would be the sum of the separate pulls. But in that case the direction of the fall of the tree might bring it on top of them.

If, however, they were to pull in different directions, as suggested in Fig. 51, they know from experience that the tree would fall somewhere between the lines of action of the forces exerted by them.

Clearly this could have been accomplished by a single

force which would have the same effect as the two forces, though not equal to their sum, and acting somewhere between. This force would be the **resultant** of the two forces.

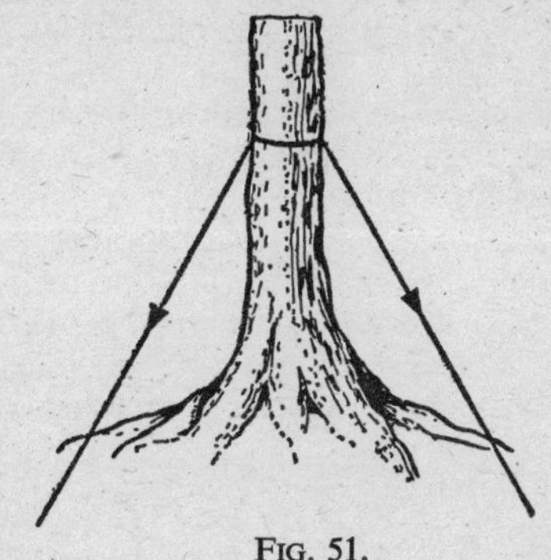

FIG. 51.

Thus we want to discover how to obtain both the magnitude and direction of the resultant of two such forces, acting in different directions on a body.

47. Resultant of two forces acting at a point

The method by which the resultant can be obtained is best demonstrated, at this stage, by experiments. The principle to be adopted will be the same as that which was used for obtaining the resultant of parallel forces. We will find what force will produce equilibrium when two forces are acting at a point. This must be equal and opposite to the resultant which can replace the two forces.

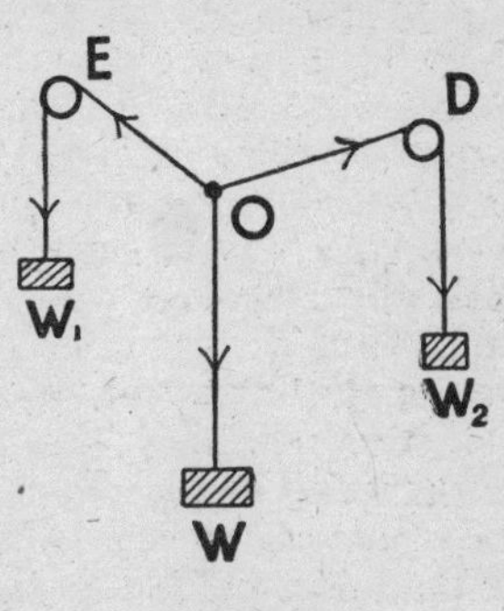

FIG. 52.

Experiment. The actual apparatus employed is shown in Fig. 52. Three cords are knotted together at a point. Two of them are passed over smooth pulleys and attached to different weights. A sufficient weight is attached to the third string to keep the others in equilibrium.

We have seen that if a string passes over a smooth pulley only the direction of the tension in the string is altered, not the magnitude of it. If the suspended weights are W_1 and W_2, then forces equal to these act along the strings. Since there is equilibrium, the tension in these strings must be balanced by the weight, called W, hung on the third string.

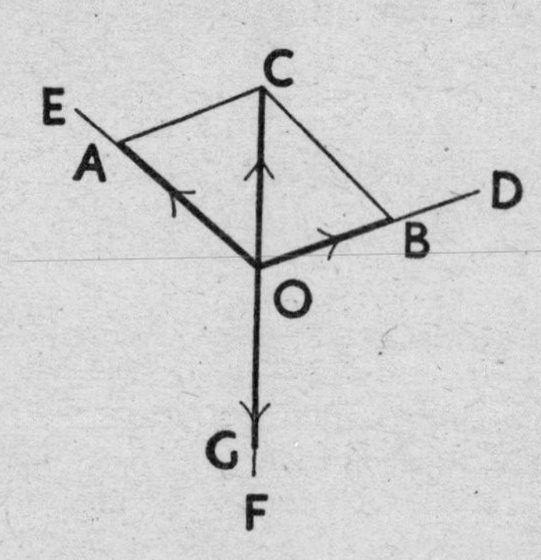

FIG. 53.

Since W is equal and opposite to the combined effects of W_1 and W_2, it must be equal and opposite to their resultant, and its line of action is that of the third string.

We now draw on the surface of the board, or on paper placed on it, straight lines corresponding to the strings *OF*, *OD*, *OE*, shown in Fig. 53.

Choosing a suitable scale:

Along *OD* mark off *OB* to represent the force acting along *OD*—i.e. W_2.
Along *OE* mark off *OA* to represent the force acting along *OE*—i.e. W_1.
Along *OF* mark off *OG* to represent the third force W, acting along *OF*.
Draw *AC* parallel to *OB* and *BC* parallel to *OA*.
Then *OACB* is a parallelogram.

1. **If *FO* is produced it will be found to pass through *C*,** so that *OC* is a diagonal of the parallelogram.

2. **Now measure *OC*. It will be found to be equal to *OG*.**

∴ **on the scale in which *OA* represents W_2, and *OB* represents W_1, *OC* must represent *W*** in magnitude.

But *W* is the force which is in equilibrium with W_1 and W_2.

∴ just as the force *W* represented by *OG*, acting downwards, maintains equilibrium, **the force represented by *OC*, acting in the opposite direction, must be the resultant of W_1 and W_2.**

∴ the diagonal of the parallelogram *OACB* represents in magnitude and direction the resultant of the forces represented by *OA* and *OB*.

This experiment should be repeated with different values for W_1 and W_2, and consequently different values of *W*.

In every case, making allowances for slight errors, you will come to the same conclusion—viz. that:

If two forces, represented by *OA* and *OB*, act at a point *O*, and include between them the angle *AOB*, then if the parallelogram *OACB* be completed, the diagonal of this parallelogram which passes through *O* represents in magnitude and direction the resultant of the forces. This is the theorem known as the Parallelogram of Forces. It may be defined thus:

Parallelogram of Forces

If two forces acting at a point are represented in magnitude and direction by two adjacent sides of a parallelogram, the diagonal of the parallelogram which passes through the point will represent their resultant in magnitude and direction.

48. Forces acting at a point

The definition of parallelogram of forces refers to two forces acting at a point; at times we speak of forces acting "on a body". In a strict sense a force cannot act "at a point" if a "point" is used with a geometrical meaning. What is implied is that the lines of action of the forces meet at a point—that is, they are concurrent. This is a necessary hypothesis of the theorem. The forces act on a body, but the point of intersection of their lines of action does not necessarily lie in the body itself.

If, for example, we consider the case of two ropes attached to a tree (§ 46), and by means of which forces act on the tree, the ropes themselves will not meet, but the lines of action of the forces exerted by means of them must meet if the parallelogram of forces is to apply to them.

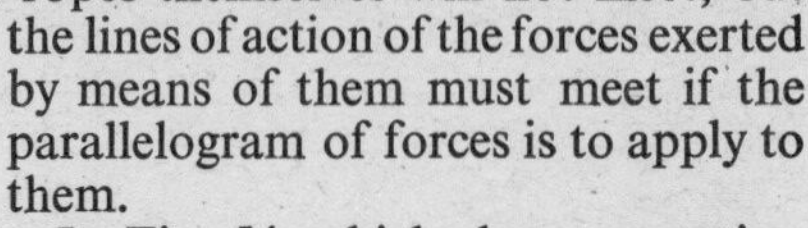

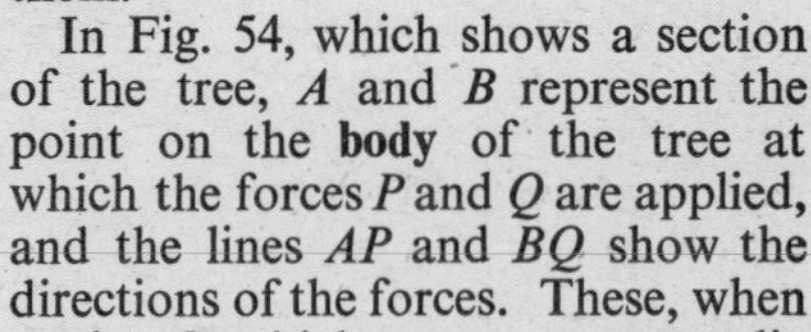

FIG. 54.

In Fig. 54, which shows a section of the tree, *A* and *B* represent the point on the **body** of the tree at which the forces *P* and *Q* are applied, and the lines *AP* and *BQ* show the directions of the forces. These, when produced, meet at a point *O*, which may or may not lie within the tree.

Nevertheless if we desire to construct a parallelogram for the forces *P* and *Q*, so that their resultant may be found, the point at which the sides of the parallelogram, which represent the forces meet, would represent the point *O*.

A force may be considered as acting at any point on its line of action. Consequently the force *P* may be considered as acting anywhere along *OA*. Similarly, *Q* may be considered as acting at any point on *OB*. Hence the two forces may be considered as acting at *O*.

49. To calculate the resultant of two forces

The problem of calculating the magnitude and direction of the resultant by means of the parallelogram of forces may be solved either by drawing or by calculation:

(1) The **drawing method** needs no explanation. This method must be used by students who have an insufficient knowledge of trigonometry. It is a useful, and sometimes an essential method, and with great care a fair degree of accuracy may be reached.

(2) The **method of calculation.** For this a working knowledge of trigonometry is essential. The student who is weak in the subject can consult the companion book to this volume, *Teach Yourself Trigonometry*, and when necessary, in the work which follows, references will be given to the appropriate pages in that book.

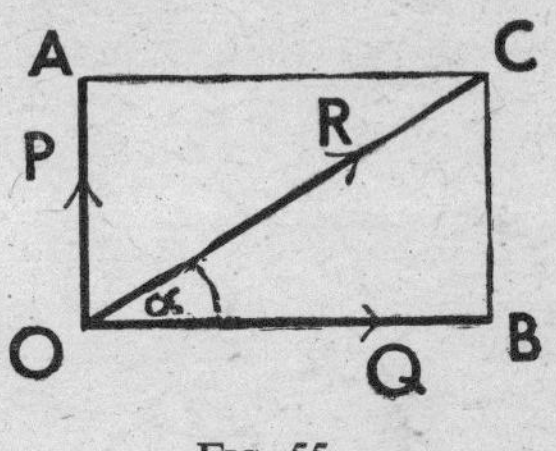

FIG. 55.

We will proceed to develop the method of calculation. Two cases may occur.

A. When the angle between the forces is a right angle—that is, the parallelogram is a rectangle.

In Fig. 55, OA, OB represent in magnitude and direction the forces P and Q. The rectangle $OACB$ is completed. Then the **diagonal through** O—i.e. OC—represents the resultant, R, of the two forces.

Let α be the angle between the force Q and the resultant. Then $90° - \alpha$ is the angle between R and P.

We require to find R and α.

(For methods employed see *Trigonometry*, § 61.)

(1) To find R:

OBC being a right-angled triangle

$$OC = \sqrt{OB^2 + BC^2}.$$

$$\therefore \quad R = \sqrt{P^2 + Q^2}.$$

(2) To find α:

$$\tan \alpha = \frac{CB}{OB} = \frac{P}{Q},$$

whence α is determined.

To find forces P and Q when R and α are known:

From ΔOBC

$$P = R \sin \alpha$$
$$Q = R \cos \alpha.$$

B. When the angle between the forces is not a right angle.

In Fig. 56, OA, OB represent forces P and Q, and θ is the angle between them.

The parallelogram $OACB$ is constructed.

Then R represents the resultant of P and Q.

Produce OB to meet perpendicular from C at D.

Since OA and BC are parallel, $\angle CBD = \theta$.

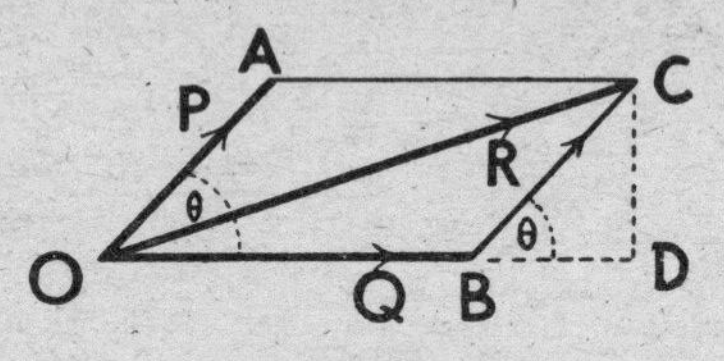

FIG. 56.

Also $BC = OA$.

$\therefore$ BC represents P.

(1) To find R:

The sides OB and BC of ΔOBC are known, since they represent the magnitudes of the forces P and Q.

The connection between the sides of the triangle and the angle OBC is given by the formula

$$OC^2 = OB^2 + BC^2 - 2OB \, . \, BC \cos OBC$$

(*Trigonometry*, § 91).

But $\angle OBC = 180° - \theta$.

$\therefore$ $\cos OBC = -\cos\theta$

(*Trigonometry*, § 70).

Substituting

$$\begin{aligned} OC^2 &= OB^2 + BC^2 - 2OB \, . \, BC(-\cos\theta) \\ &= OB^2 + BC^2 + 2OB \, . \, BC\cos\theta. \end{aligned}$$

$\therefore$ on substitution

$$\mathbf{R^2 = P^2 + Q^2 + 2PQ\cos\theta} \quad . \quad . \quad (1)$$

From this R can be found when P, Q, and θ are known.

By a slight modification of the figure and proof, the same result may be obtained when $\angle AOB$ is greater than a right angle.

(2) To find α, the angle between Q and R:

From Fig. 56 we see

$$BD = BC \cos \theta$$
$$= P \cos \theta.$$

Similarly $$CD = P \sin \theta.$$

Now $$\tan \alpha = \frac{CD}{OD} = \frac{CD}{OB + BD}.$$

Substituting values found above

$$\tan \alpha = \frac{P \sin \theta}{Q + P \cos \theta} \quad . \quad . \quad . \quad (2)$$

From this formula α may be calculated.

Also $$\angle AOC = \theta - \alpha.$$

(3) To find θ when P, Q, and R are known:

These may be found by transforming formula (1) above, viz.:

$$R^2 = P^2 + Q^2 + 2PQ \cos \theta.$$

From this

$$\cos \theta = \frac{R^2 - P^2 - Q^2}{2PQ} \quad . \quad . \quad . \quad (3)$$

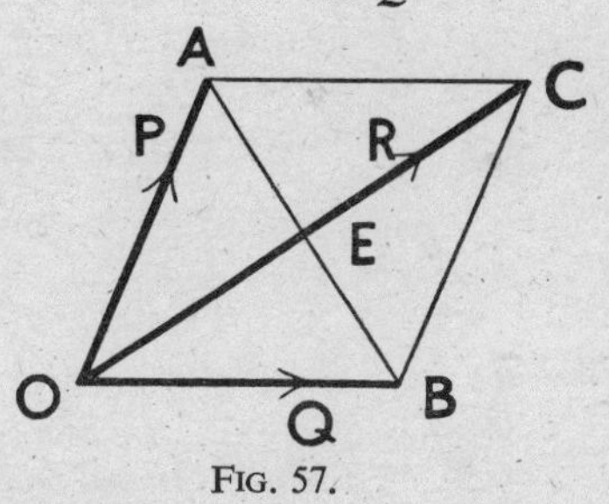

FIG. 57.

(4) To find R when P and Q are equal:

The parallelogram which was described in Fig. 56 now becomes a rhombus: therefore,

(*a*) $\angle \theta$ is bisected and R makes an angle $\frac{\theta}{2}$ with both P and Q.

(*b*) The diagonals bisect each other at right angles.

$\therefore$ Δs OEA, OEB are right-angled.

$$\therefore \quad \frac{OE}{OB} = \cos EOB$$

$$= \cos \frac{\theta}{2}.$$

$$\therefore \quad OE = OB \cos \frac{\theta}{2}.$$

But $OC = 2 \times OE$.

$\therefore$ $OC = 2OB \,.\, \cos \frac{\theta}{2}$ on substitution for OE

or $\quad R^2 = 2Q \cos \frac{\theta}{2}$ . . . . (4)

50. Summary of formulæ

The above formulæ are used so frequently in solving the problems that they are collected for future reference.

(1) **To find R:**

$$R = P^2 + Q^2 + 2P \,.\, Q \,.\, \cos \theta.$$

(2) **To find α:**

$$\tan \alpha = \frac{P \sin \theta}{Q + P \cos \theta}.$$

(3) **To find θ:**

$$\cos \theta = \frac{R^2 - P^2 - Q^2}{2PQ}.$$

(4) **To find R when P and Q are equal:**

$$R = 2Q \cos \frac{\theta}{2}.$$

51. Resultant of a number of forces acting at a point

When more than two forces act at a point the resultant of them all can be found by repeated application of the parallelogram law for two forces. We proceed as follows:

The resultant of two forces is found. Let it be R_1.

Next the resultant of R_1 and a third force is found. Let it be R_2.

Similarly, the resultant of R_2 and a fourth force is found. This process is repeated until the resultant of all the forces has been found.

In practice this method is often long and tedious, but by careful drawing a practical solution can be obtained.

An alternative method will be given later.

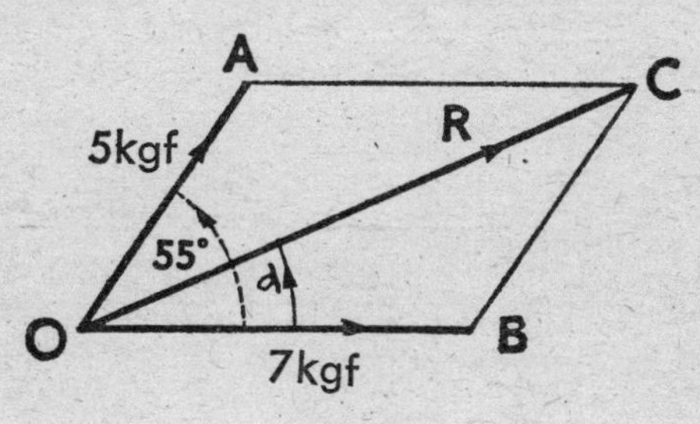

FIG. 58.

52. Worked Examples

Example 1. *Forces of 7 kgf and 5 kgf act on a body, and the angle between them is 55°. Find their resultant and the angles between it and the two forces.*

Fig. 58 represents the given forces

Required to find R and α.

(1) **To find R:**

Using formula (1)

$$R^2 = P^2 + Q^2 + 2PQ \cos \theta,$$

and substituting the given values of P, Q, and θ.

$$\begin{aligned} R^2 &= 5^2 + 7^2 + 2 \cdot 5 \cdot 7 \cos 55^\circ \\ &= 74 + 70 \times 0{\cdot}5736 \\ &= 114{\cdot}15. \\ \therefore \quad R &= \sqrt{114{\cdot}15} \\ &= \mathbf{10{\cdot}7\ kgf\ approx.} \end{aligned}$$

(2) **To find α:**

Using formula (2)

$$\tan\alpha = \frac{P\sin\theta}{Q + P\cos\theta}.$$

We have on substitution:

$$\begin{aligned}
\tan\alpha &= \frac{5\sin 55^\circ}{7 + 5\cos 55^\circ} \\
&= \frac{5\times 0{\cdot}8192}{7 + 5\times 0{\cdot}5736} \\
&= \frac{4{\cdot}0960}{9{\cdot}8680}. \\
\therefore \quad \log\tan\alpha &= \log 4{\cdot}0960 - \log 9{\cdot}8680 \\
&= 0{\cdot}6123 - 0{\cdot}9943 \\
&= \bar{1}{\cdot}6180 \\
&= \log\tan 22^\circ\,32'. \\
\therefore \quad \alpha &= 22^\circ\,32'. \\
\text{Also} \quad \angle AOC &= \angle AOB - \alpha \\
&= 55^\circ - 22^\circ\,32' \\
&= 32^\circ\,28'.
\end{aligned}$$

This could be checked by calculating $\angle AOC$ separately, using formula (2).

Example 2. *Two forces, P and Q, of* 10 *kgf and* 14 *kgf respectively, have a resultant of* 22 *kgf. What is the angle between P and Q?*

If θ be the angle between P and Q,

Using formula (3)

$$\cos\theta = \frac{R^2 - P^2 - Q^2}{2PQ}.$$

Substituting given values

$$\begin{aligned}
\cos\theta &= \frac{22^2 - 10^2 - 14^2}{2\,.\,10\,.\,14} \\
&= \frac{188}{280} = 0{\cdot}6714. \\
\therefore \quad \theta &= 47^\circ\,50'.
\end{aligned}$$

53. Moments of intersecting forces

We have seen (§ 37) that the moments of a number of parallel forces about a point in their plane is equal to the moment of their resultant about the point.

We shall show that this is also true for intersecting forces. First we will consider a graphic method of representing moments.

Graphical representation of moments

Let AB, Fig. 59, represent a force F in magnitude and direction.

Let O be any point.

Join OA, OB.

Draw OD perpendicular to AB.

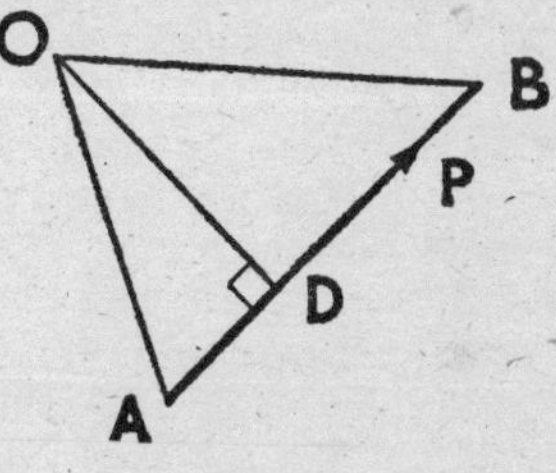

FIG. 59.

Since AB represents the force F, the turning moment of F about O is equal to the magnitude of the force multiplied by its distance from O,

i.e. $\qquad$ moment of $F = AB \times OD$.

But area of $\Delta OAB = \frac{1}{2}(AB \times OD)$.

$\therefore$ **Area of $\Delta OAB = \frac{1}{2} \times$ (moment of F about O).**

This method will be employed in the following demonstration.

54. The algebraic sum of the moments of two intersecting forces about a point in their plane is equal to the moment of their resultant about the same point.

Let OA and OB (Fig. 60) represent in direction two forces F_1 and F_2 intersecting at O.

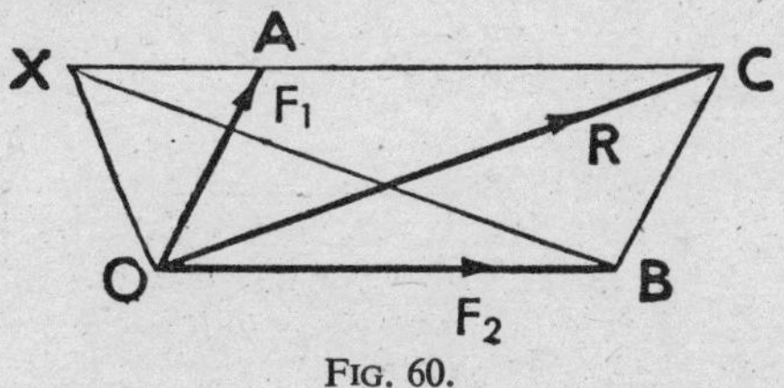

FIG. 60.

Let X be any point in their plane.

Draw XAC parallel to OB and meeting the line of force F_1 in A.

Let the scale with which the magnitudes of the forces F_1 and F_2 are drawn be such that OA represents F_1 and OB represents F_2.

Draw BC parallel to OA to meet XAC in C.

Then $OACB$ is a parallelogram and OC, its diagonal, represents R, the resultant of F_1 and F_2.

Taking moments about X.

Moment of F_1 is represented by $2 \times$ area of ΔOAX.
Moment of F_2 is represented by $2 \times$ area of ΔOBX.
Moment of R is represented by $2 \times$ area of ΔOCX.

But $\Delta OCX = \Delta OAX \times \Delta OAC$

and $\Delta OAC = \Delta OBC$

$= \Delta OBX$ (Δs on same base and between same parallels).

$\therefore \quad \Delta OCX = \Delta OAX + \Delta OBX$

and $2 \times \Delta OCX = 2 \times \Delta OAX + 2 \times \Delta OBX$,

i.e. **moment of R about X = moment of F_1 about X. + moment of F_2 about X.**

Note. If the point X be taken on the line of action of R the **moment of R vanishes.**

Consequently the sum of the moments of F_1 and F_2 about the same point also vanishes.

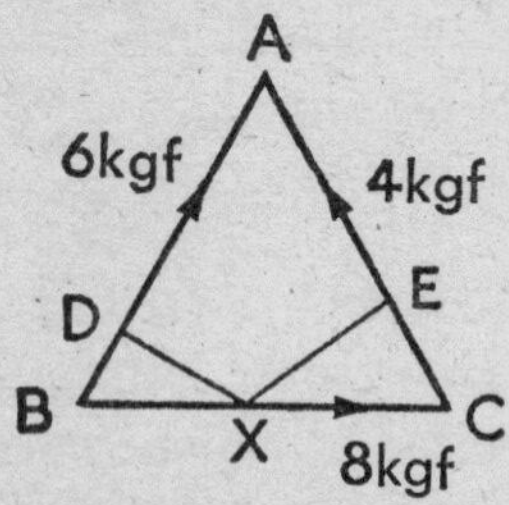

Fig. 61.

55. Worked example

Forces of 4 kgf, 6 kgf and 8 kgf act along the sides of an equilateral triangle of 10 cm side, in directions shown in Fig. 61. Find the line of action of their resultant.

Let the line of action of resultant cut BX in X.

Draw XD and XE perpendicular to AB and AC.

Since X lies on the line of action of the resultant moments of the forces about X must vanish (§ 54, note).

Also as X lies on BC, the moment of the force along BC also vanishes.

$\therefore$ the sum of the moments of the forces along BA and CA must vanish.

$\therefore$ moment of force along BA = moment of force along CA.

$\therefore$ $6 \times DX = 4 \times EX.$

But $DX = BX \sin 60$

and $EX = CX \sin 60.$

$\therefore$ $6 \times BX \sin 60 = 4 \times CX \sin 60.$

$\therefore$ $6BX = 4CX.$

$\therefore$ $\dfrac{BX}{CX} = \dfrac{4}{6}.$

$\therefore$ X must divide BC in the ratio 4:6.

Similarly, if a point Y be taken where the line of action of the resultant cuts AC, it will be found by the same reasoning that Y divides AC in the ratio 8:6.

$\therefore$ the straight line joining the points X and Y will be the line of action of the resultant.

Exercise 6

1. F' and F'' are two forces and F is their resultant. The angle between the two forces is θ and the angle between F and F'' is α.

(a) Find F when $F' = 10$, $F'' = 24$, $\theta = 90°$.
(b) Find F and α when $F' = 16{\cdot}5$, $F'' = 22$, $\theta = 90°$.
(c) Find F when $F' = 5$, $F'' = 8$, $\theta = 42°$.
(d) Find F and α when $F' = 4$, $F'' = 6$, $\theta = 60°$.
(e) Find θ when $F' = 36$, $F'' = 60$, $F = 80$.
(f) Find θ and α when $F' = 20$, $F'' = 16$, $F = 28$.

2. Fig. 62 shows a weight of 5 kgf suspended by a string from B, which is drawn aside by a force acting along a string OA, knotted at O and stretched horizontally. It is thus held in equilibrium, the force acting along OA being 3 kgf. Find:

(1) The pressure exerted at B.
(2) The angle that OB makes with the vertical.

3. Fig. 63 shows a weight of 10 kgf sustained by a rope passing over a smooth pulley, and the two parts of the

rope are inclined at an angle of θ. Find the magnitude and direction of the resultant pressure on the pulley for the following values of θ:

(1) 0°; (2) 30°; (3) 45°; (4) 50°; (5) 60°; (6) 90°.

Plot your results on squared paper.

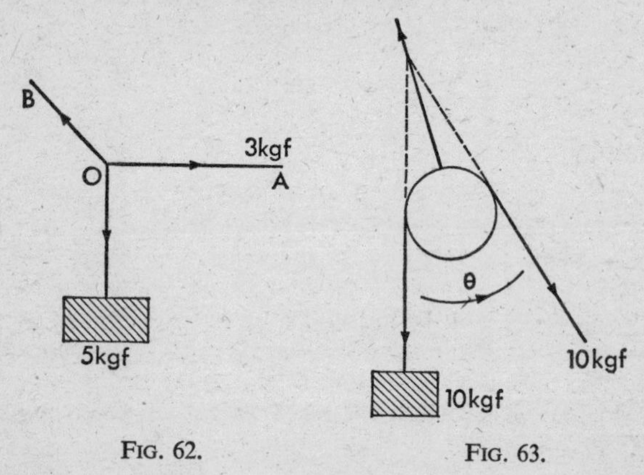

FIG. 62. FIG. 63.

4. The resultant of two forces is 16 kgf. One of the forces is 14 kgf and makes an angle of 45° with the resultant. What is the magnitude of the other force?

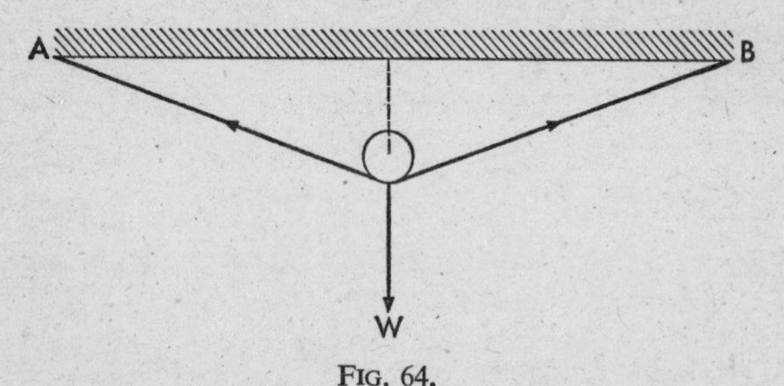

FIG. 64.

5. Two forces of 8 kgf and 7 kgf have a resultant of 13 kgf. What is the angle between the two forces?

6. Two forces of 14 gf and 10 gf have a resultant of 12·5 gf. Find the angle between them.

7. A wire suspended from two points A and B (*see* Fig. 64) supports at its centre a small smooth pulley from which hangs a weight of W kgf. The angle between the two parts of the wire is 140° and the force exerted in each part of the wire is 150 kgf. What is the value of the load, W?

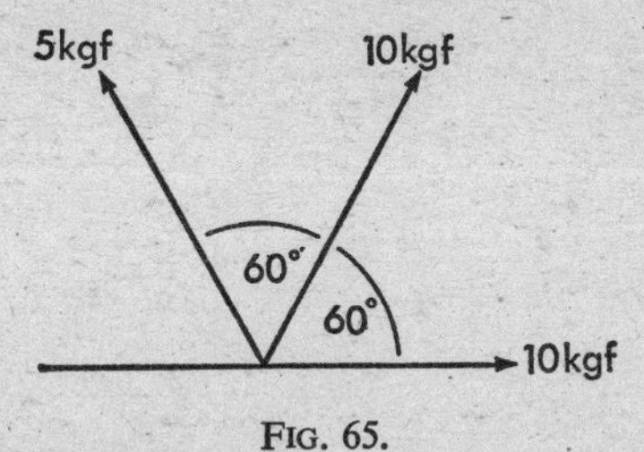

FIG. 65.

8. Two forces of 10 kgf and a force of 5 kgf act at a point, and the angle between each pair is 60° as shown in Fig. 65. Find the resultant of the three forces.

9. Forces act along the sides of an equilateral triangle ABC as follows:

0·9 kgf from B to C;
0·6 kgf from C to A;
0·3 kgf from B to A.

In what ratio are the sides BC and CA cut by the resultant?

CHAPTER V

COMPONENTS OF A FORCE; RESOLVED PARTS OF A FORCE

56. Components of a force

We have seen that two forces whose lines of action intersect can be replaced by a resultant, the magnitude and direction of which can be found by the application of the theorem the Parallelogram of Forces.

The two forces are called the **components** of the resultant force. The problem now to be considered is how these components are to be found when the resultant is known.

When applying the theorem of the Parallelogram of Forces to find the resultant of two known forces, there is only one solution, since but one diagonal of the parallelogram can pass through the intersection of two adjacent sides. But in the converse problem a given straight line may be a diagonal of an infinite number of parallelograms; consequently it may be the resultant of an infinite number of pairs of forces, differing in magnitude and direction.

When, however, the direction of the components is known, there is only one solution. In these circumstances the components may be found readily, either by drawing or by using the appropriate formula of the preceding chapter.

57. Resolving a force

The case when the two unknown components are at right angles is much the most important from a practical point of view. The components thus obtained are called **the resolved parts in the given direction** and the process of obtaining them is called **"resolving the force".**

Thus in Fig. 66 the force R, acting along and represented by OC, can be resolved into two forces, F_1 and F_2, one of which makes an angle θ with R, and the other which makes an angle of $90° - \theta$ with R.

F_1 and F_2 are the resolved parts of R in the direction θ and $90° - \theta$ with R.

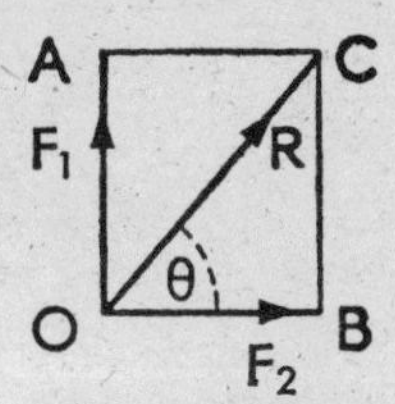

FIG. 66.

58. To find the resolved parts of a force

To determine the values of the resolved parts, complete the rectangle $ABCD$ (Fig. 66) with OA representing F_1 and OB representing F_2.

Then $$\frac{OB}{OC} = \cos\theta.$$

$\therefore$ $$OB = OC\cos\theta.$$

But OB represents F_2 in magnitude and direction,
and OC represents R in magnitude and direction,

$\therefore$ $$\mathbf{F_2 = R\cos\theta.}$$

Also since $$BC = OA,$$

and $$\frac{BC}{OC} = \sin 0,$$

$\therefore$ $$OA = OC\sin\theta,$$

i.e. $$\mathbf{F_1 = R\sin\theta.}$$

$\therefore$ **Resolved part of R along $OB = R\cos\theta$.**
and **resolved part of R along $OA = R\sin\theta$.**

Since a force has no effect in a direction at right angles to its own direction,

$R\cos\theta$ is the **total effect of R in the direction OB,**
and $R\sin\theta$ is the **total effect of R in the direction OA.**

59. Illustrative example

The following example will serve as a practical illustration of this principle. Fig. 67 represents a barge, LO, to which is attached at O a rope by means of which a horse at A, moving along the tow-path of a canal, drags the barge through the water. The pull of the horse, from the necessities of the case, cannot be a direct one, and consequently a certain amount of the effort is lost.

The useful part of the pull is in the direction indicated by *OB*.

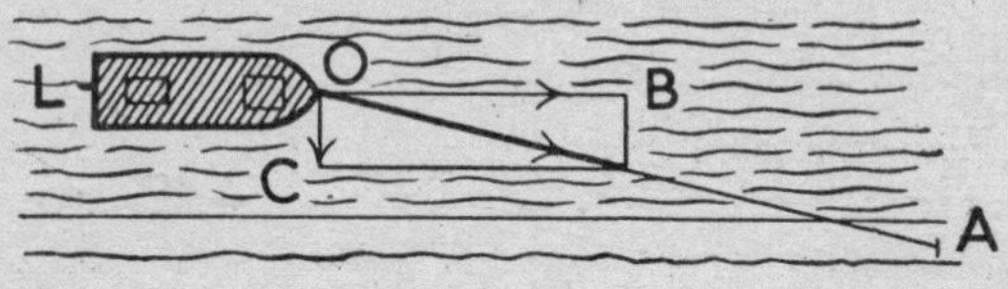

FIG. 67.

This is the resolved part of the pull in the direction in which the barge is being towed.

Let θ be the angle between *OA* and *OB*.

Resolving the pull (*F*) of the horse in the direction shown by *OB* and also at right angles to it, along *OC*, we see, as shown in § 58, that *F* is resolved into

$$F \cos \theta \text{ acting along } OB,$$
$$F \sin \theta \text{ acting along } OC.$$

Thus the effective pull of the horse is $F \cos \theta$, and $F \sin \theta$ represents the lateral pull towards the bank which is neutralised by the set of the rudder.

The useful part of the pull of the horse thus depends on $\cos \theta$. Now we know from Trigonometry that $\cos \theta$ is always less than unity; consequently $F \cos \theta$ is always less than *F*. Further, as an angle increases, the cosine decreases. Therefore, as the pull is more effective when $\cos \theta$ is large, θ should be as small as possible.

It is therefore advantageous, in general, that the rope should be long, because then θ is small. If it be too long, however, other considerations affect the efficiency of the pull.

60. Forces acting on a body lying on a slope

A smooth ball placed on a smooth horizontal board will rest on it in neutral equilibrium. But if the board be tilted, however slightly, the ball will begin to move.

Gravity is the only force acting which would produce

motion. This force acting vertically has no resolved part in a horizontal direction. There is therefore no motion while the board is horizontal. Since it moves when the board is sloped, this must be due to the resolved part of the force of gravity along the sloping surface.

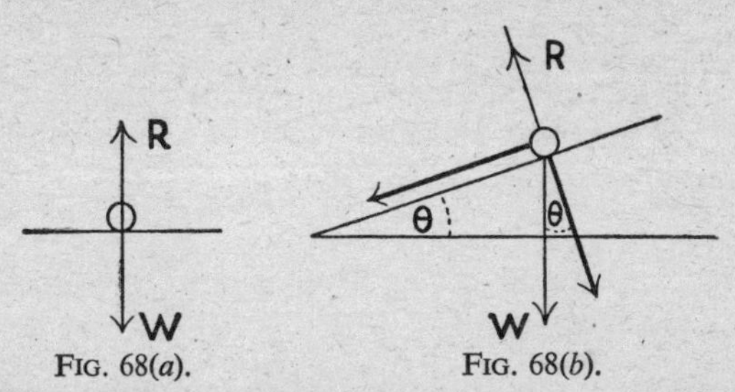

FIG. 68(*a*). FIG. 68(*b*).

In Fig. 68(*a*) the diagram shows the forces acting on the ball when it is at rest on the horizontal surface. These are:

(1) The weight, i.e. the force of gravity.

(2) The reaction of the board equal and opposite to the forces of gravity, thus producing equilibrium.

In Fig. 68(*b*) let θ be the angle of slope of the board. Then the angle between the vertical and the straight line perpendicular to the board is also θ.

$\therefore$ The force W can be resolved into

(1) **$W \cos \theta$, acting at right angles to the surface of the board.**

(2) **$W \sin \theta$, acting along the board.**

It is the resolved part acting down the surface of the board, viz. $W \sin \theta$, which causes the ball to move **down** the plane. If there is to be equilibrium in the new position a force equal and opposite to this must act on the body in a direction **up** the plane.

The reaction of the board will be equal and opposite to the component $W \cos \theta$, acting at right angles to the board. This reaction is therefore less when the board is tilted than when it is horizontal, and it decreases as θ is increased, since $\cos \theta$ decreases. At the same time $W \sin \theta$ increases as θ increases.

61. Application of moments to resolved parts

It was shown in § 54 that the sum of the moments of two intersecting forces about any point in their plane is equal to the moment of their resultant about the same point.

When a force is resolved into two parts, it becomes the resultant of the two components; therefore **the moment of the force about any point is equal to the sum of the moments of its resolved parts.**

This principle is useful in solving problems. An example follows:

62. Worked example

A uniform beam, OB, 4 *m long and weighing* 150 *kgf, is hinged at O* (*Fig.* 69). *It is kept in equilibrium in a horizontal position by a wire rope fixed at a point A*

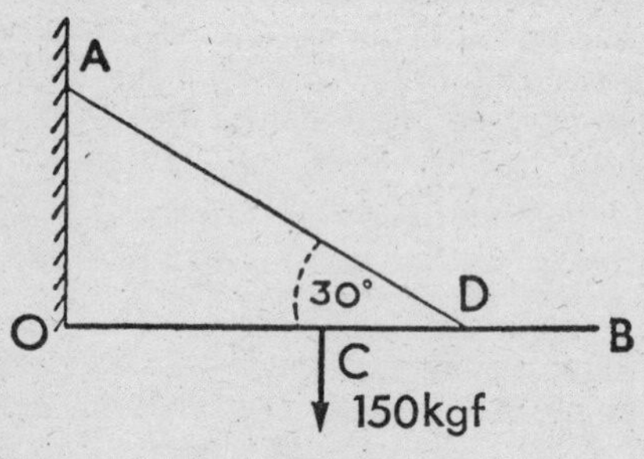

FIG. 69.

vertically above O, and attached to the beam at a point D, 3 *m from O. If the wire makes an angle of* 30° *with the horizontal, what is the tension in it?*

The forces acting on the beam are:

(1) The **tension,** T, in the wire.
(2) **Weight** of beam at mid point, C.
(3) **Reaction** at O.

If we take moments about O, that of the reaction, which is unknown, is eliminated.

If T be resolved into its vertical and horizontal components, $T \sin 30°$ and $T \cos 30°$, the latter will pass through O, and therefore has no moment about it.

We are therefore left with:

(1) **Vertical component of T,** viz. $T \sin 30°$, acting at D.

(2) **Weight of beam,** 150 kgf, acting at C.

These have opposite turning moments about O, and as there is equilibrium they must be equal.

$$\therefore \quad T \sin 30° \times 3 = 150 \times 2$$

$$T = \frac{150 \times 2}{3 \sin 30°}$$

$$= \mathbf{200\ kgf.}$$

Exercise 7

1. A force of 16 kgf acts along a straight line OA (Fig. 70) which is inclined at an angle of 30° with OX. Find its resolved parts along OX and OY, which is perpendicular to OX.

2. Find the resolved parts along OX and OY (Fig. 70) when the force along OA is 12 kgf and the angle AOX is 25°.

3. Resolve a force, F, of 8 kgf, into two components at right angles and such that one of them makes an angle of 55° with the direction of F.

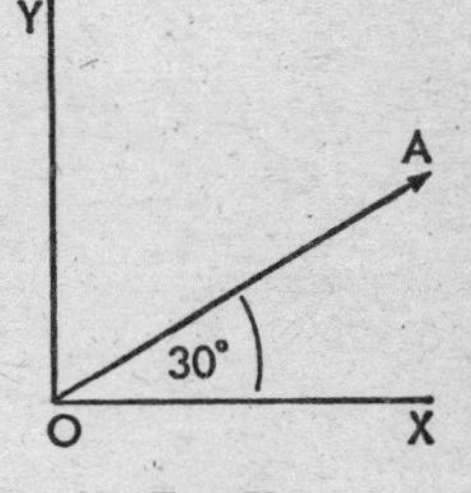

FIG. 70.

4. Two tugs are towing a ship. The rope from one of them is inclined at 15° to the path of the ship, and the other at 20° (Fig. 71). If each tug exerts a force equal to 10 tf, what is the effective force in the direction of the ship's motion?

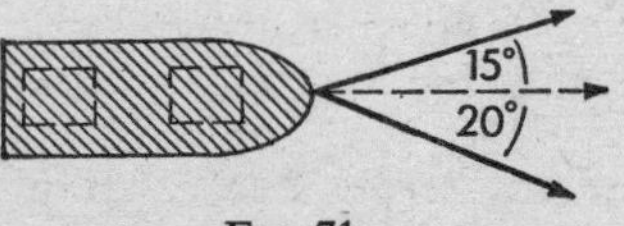

FIG. 71.

5. Find the vertical and horizontal resolved parts of the following forces:

(1) A force of 20 kgf acting at 40° with the horizontal.
(2) 250 gf acting at 30° with the horizontal.

6. A man, moving along the bank, pulls a boat along a river. The rope by which he pulls the boat is inclined at 20° to the direction of the boat's motion. If the man exerts a force of 100 kgf, what is the effective force in the direction of the boat's motion?

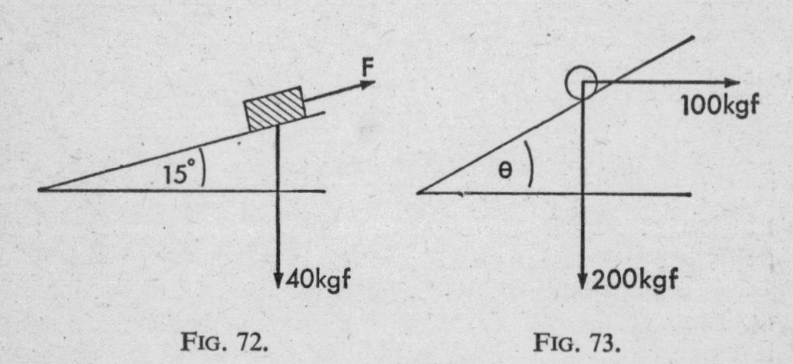

FIG. 72. FIG. 73.

7. A mass of 40 kg lies on a smooth inclined surface which makes an angle of 15° with the horizontal (Fig. 72). If it is just kept in position by a force of F kgf, acting parallel to the sloping surface, what is the magnitude of F?

8. A force of 100 kgf acting horizontally maintains in equilibrium a cylindrical roller weighing 200 kgf on a smooth surface inclined at an angle of θ to the horizontal (Fig. 73). Find the value of θ.

If θ were 45°, what horizontal force would be necessary to maintain the 200 kg in equilibrium?

9. A stay wire is fastened to a peg embedded in the ground, with which the wire makes an angle of 50° (Fig. 74). If the pull along the wire is 40 kgf, what vertical force is exerted on the peg to tend to pull it out of the ground?

10. A 100 kg mass rests on a smooth surface inclined at an angle of 15° to the horizontal. It is kept in equilibrium by a force F which makes an angle of 20° with the inclined surface (Fig. 75). What is the magnitude of F?

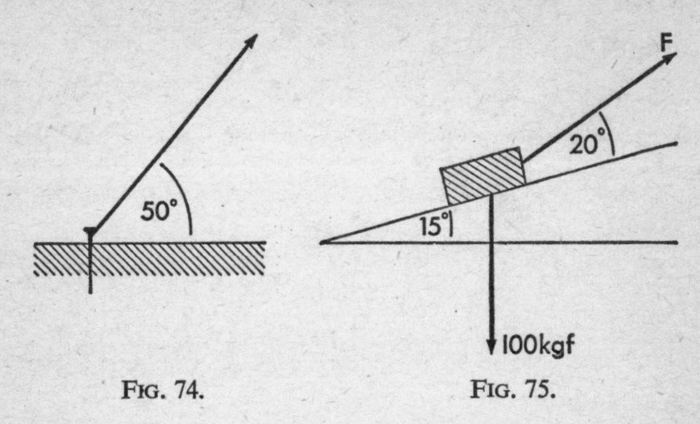

FIG. 74. FIG. 75.

63. Resolved parts. Resultant of forces acting at a point

When a number of forces act at a point their resultant can be obtained by resolving the forces along two directions at right angles and then adding these components. The following example will illustrate the method.

Example. *Two forces F_1 and F_2 of 8 kgf and 10 kgf act along straight lines, the angle between which is 45°. Find their resultant.*

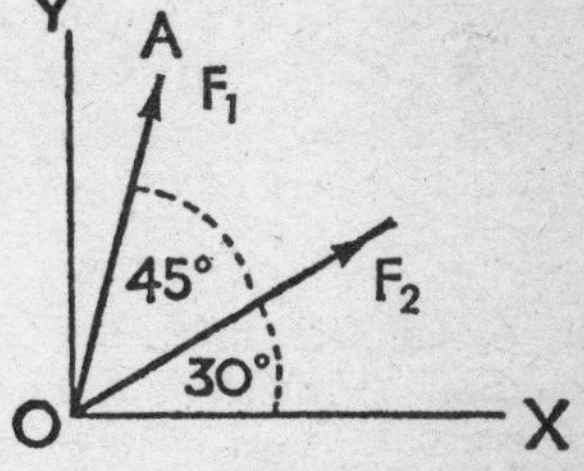

FIG. 76.

Let OA, OB (Fig. 76) be the lines of action of the forces F_1 and F_2, the angle between them being 45°.

Take two straight lines, OX, OY, at right angles, such that $\angle XOB = 30°$. Then $\angle AOY = 15°$ and $\angle AOX = 75°$.

Note. Unless straight lines at right angles are specified, OX and OY can be chosen in the way which is most suitable.

Find the resolved parts of F_1 and F_2 along OX and OY:

(1) **Resolved parts along OX.**

$$F_2 \cos 30°, F_1 \cos 75°,$$

i.e. $$10 \cos 30°, 8 \cos 75°.$$

Total $= 10 \cos 30° + 8 \cos 75° = 10{\cdot}73$ kgf (by calculation not shown).

(2) **Resolved parts along OY.**

$$F_2 \sin 30°, F_1 \sin 75°,$$

i.e. $$10 \sin 30°, 8 \sin 75°.$$

Total $= 10 \sin 30° + 8 \sin 75° = 12{\cdot}73$ kgf.

As we have seen (§ 58), these forces represent the **total** effect of the forces F_1 and F_2 in these two directions.

∴ **The resultant of 10·73 kgf along OX and 12·73 kgf along OY is the same as the resultant of F_1 and F_2** (*see* Fig. 77).

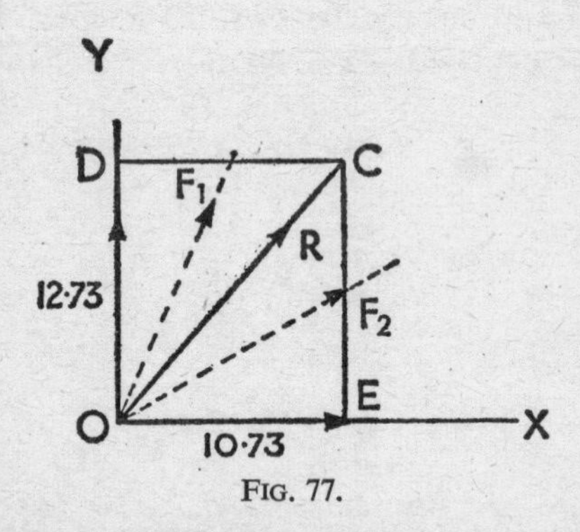

FIG. 77.

∴ if OE be measured along OX, equal to 10·73,
and OO be measured along OY, equal to 12·73,
and the rectangle $ODCE$ completed, **the diagonal of this rectangle, OC, represents the resultant.**

Let R be the resultant.

Then
$$R = \sqrt{(10{\cdot}73)^2 + (12{\cdot}73)^2}$$
$$= \mathbf{16{\cdot}7\ kgf} \text{ nearly (by calculation).}$$

If α be the angle made by R with OX,

then $$\tan \alpha = \frac{12 \cdot 73}{10 \cdot 73} = 1 \cdot 1868,$$

and $$\alpha = 49° \, 53'.$$

64. Resultant of any number of concurrent forces

This method may not appear to offer much advantage when only two forces are concerned. But it is very valuable when the number of forces is large.

Before proceeding to the general case, it is necessary to consider difficulties which arise when the angles made by the lines of action of the force with either of the axes of reference, OX and OY, as described in § 63, are greater than right angles.

To simplify the problem we will first consider the case shown in Fig. 78. A force F acts at O, and its line of action makes an angle XOB with OX, which with OY at right angles to it are the axes of reference. The angle XOB is greater than a right angle.

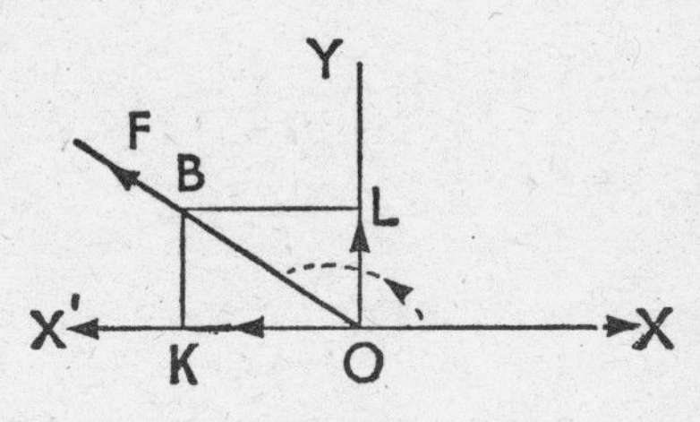

FIG. 78.

It should be remembered that, as in Trigonometry, angles are measured in an anti-clockwise direction, as indicated by the arrow-head on the arc in the diagram.

Let OB represent the magnitude of the force.

Drawing BL perpendicular to OY, OL **represents the resolved part of** F **along** OY, i.e. $F \sin BOX$.

In order to obtain the resolved part of F along OX, this line must be produced in the **opposite direction,** i.e. along OX'. Then the perpendicular BK from B

cuts off OK, which must represent the resolved part of F along OX.

$$\therefore \quad OK = F \cos BOX.$$

Two points must now be noted:

(1) The cosines of angles between 90° and 180° are negative.

$$\therefore \quad F \cos BOX \text{ is negative.}$$

(2) In accordance with the convention of positive and negative directions, when XO is produced in the **opposite** direction from O, distances measured in this direction from O to X' are considered as negative, as contrasted with distances measured from O to X which are considered positive.

$\therefore$ OK, the resolved part of F, is negative.

This is in agreement with the statement above that $F \cos BOX$ is negative, since this is founded on the same convention.

The interpretation of the negative sign in $F \cos BOX$ is that **this force, represented by OK, acts in the opposite direction** from forces which act from O towards X.

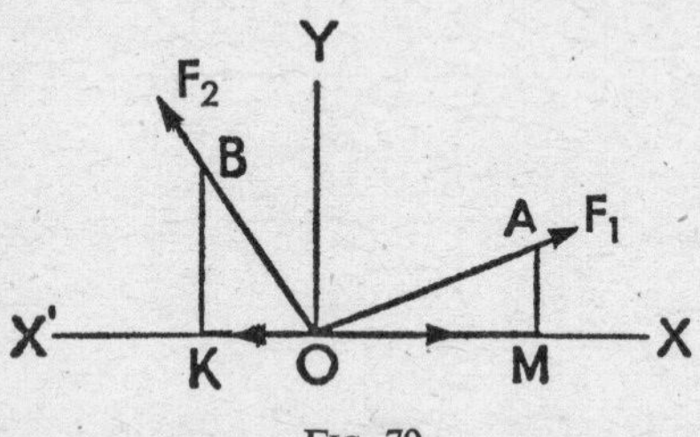

FIG. 79.

Let us now consider two forces F_1 and F_2 acting as shown in Fig. 79, in which OA, OB represent the magnitudes of these forces. Drawing AM and BK perpendicular to XOX'

OM represents resolved part of F_1 along OX.
OK represents resolved part of F_2 along OX.

∴ the algebraic sum of these resolved parts is

$$F_1 \cos AOX - F_2 \cos BOX$$
$$= OM - OK.$$

65. These principles may be extended to any number of forces which intersect at O.

To include all cases the line OY must also be produced in the opposite direction, so that the axes are drawn as shown in Fig. 80. We thus get four spaces, *I, II, III,* IV, which are called quadrants, and the line of action of any force acting at O must fall into one of these.

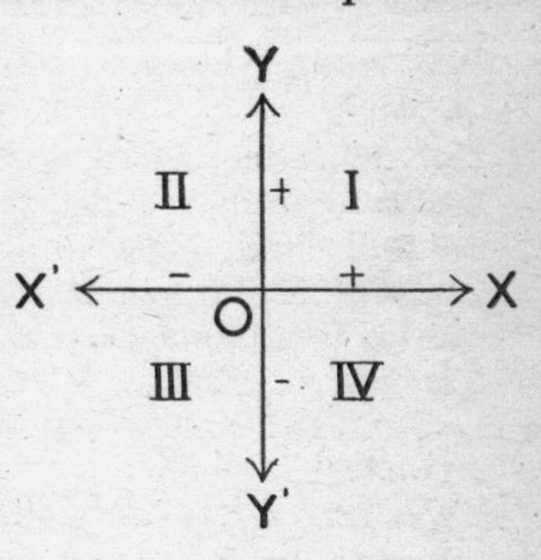

FIG. 80.

Using the convention above, measurements along OY' will be negative.

In general:

Resolved parts along OX and OY are positive.

Resolved parts along OX' and OY' are negative.

As an example, consider the cases of two forces, F_3 and F_4, falling in the IIIrd and IVth quadrants as shown in Fig. 81.

Let OE, OH represent the magnitudes of F_3 and F_4.

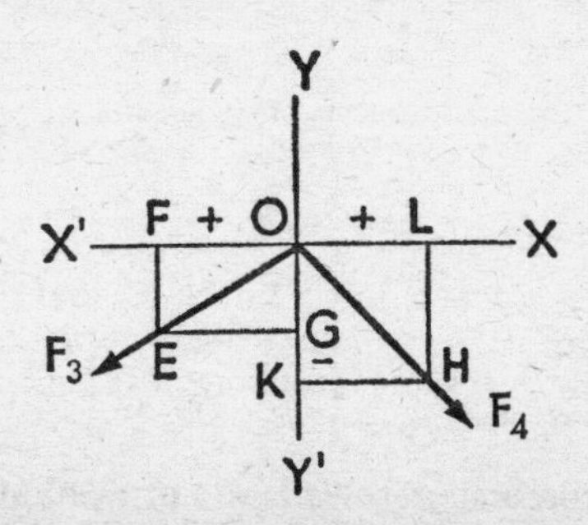

FIG. 81.

Draw perpendiculars to the axes as shown:

(1) **For the force F_3 in Quad. III:**

OF resolved part along OX' is **negative.**
OG resolved part along OY' is **negative.**

(2) **For the force F_4 in Quad. IV:**

OL resolved part along OX is **positive.**
OK resolved part along OY' is **negative.**

Considering the sum of resolved parts along each axis:

(1) **Along OX.** Sum of resolved parts $= OL - OF$.
(2) **Along OY.** Sum of resolved parts $= - OG - OK$.

The sum along OX will be positive if $OL > OF$.
The sum along OY is negative.

All this should offer no difficulty to the student whose knowledge of Trigonometry is good. A full treatment from the Trigonometrical point of view will be found in *Trigonometry*, Chapter V.

66. Worked example

Find the resultant of forces as shown in Fig. 82:

Viz. 10 *kgf acting along OA*
5 *kgf acting along OB*
10 *kgf acting along OC*
12 *kgf acting along OD*

the angles made by the lines of action of the forces with XOX' and YOY' being as shown.

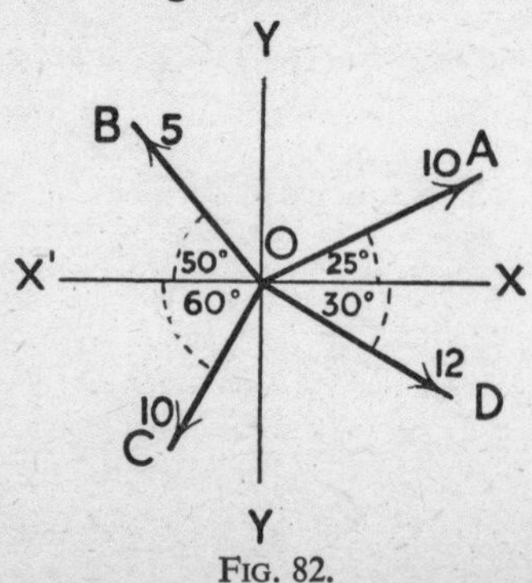

FIG. 82.

Careful arrangement of the resolved parts is important, and some such tabulation as the following is suggested:

Components along XOX'.

	Positive	Value	Negative	Value
Quad. 1	10 cos 25°	9·063		
Quad. II			5 cos 50°	3·214
Quad. III			10 cos 60°	5·000
Quad. IV	12 cos 30°	10·39		
Total	+	19·45	—	8·214

Sum = 11.24 kgf (approx.).

Components along YOY'.

	Positive	Value	Negative	Value
Quad. I	10 sin 25°	4·226		
Quad. II	5 sin 50°	3·830		
Quad. III			10 sin 60°	8·660
Quad. IV			12 sin 30°	6·000
Total	+	8·056	—	14·660

Sum = — 6·604 kgf (approx.).

The negative sign shows that the total resolved par* acts along OY'.

The forces now reduce to two, as shown in Fig. 83.

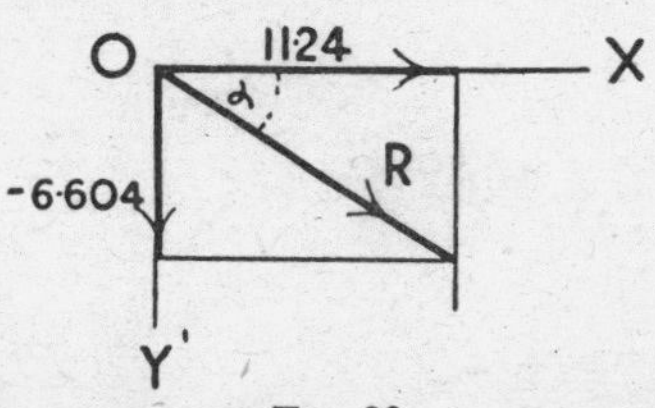

FIG. 83.

Completing the rectangle, OA represents the resultant R of the whole system.

To find R:

$$\begin{aligned} R &= \sqrt{(11\cdot24)^2 + (-6\cdot604)^2} \\ &= \sqrt{170} \\ &= \mathbf{13\ kgf\ nearly.} \end{aligned}$$

Let α be the angle made by the resultant with OX. Then, taking numerical values only,

$$\tan \alpha = \frac{6\cdot604}{11\cdot24} = 0\cdot589 \text{ approx.}$$

$$\therefore \qquad \alpha = 30° \, 31'.$$

The negative sign, if retained, would show that the angle is in the 4th quadrant (*Trigonometry*, p. 160).

67. Equilibrium of forces acting at a point

If the resultant of forces acting at a point is zero, the forces are in equilibrium.

Consequently, **if the sum of the resolved parts in two directions is zero, the resultant must be zero and the forces are in equilibrium.**

Exercise 8

1. Using the methods of resolved parts, find the resultant of forces F_1 and F_2 as shown in Fig. 84, where:

F_1 is 8 kgf and makes 45° with OX;
F_2 is 10 kgf and makes 60° with OX.

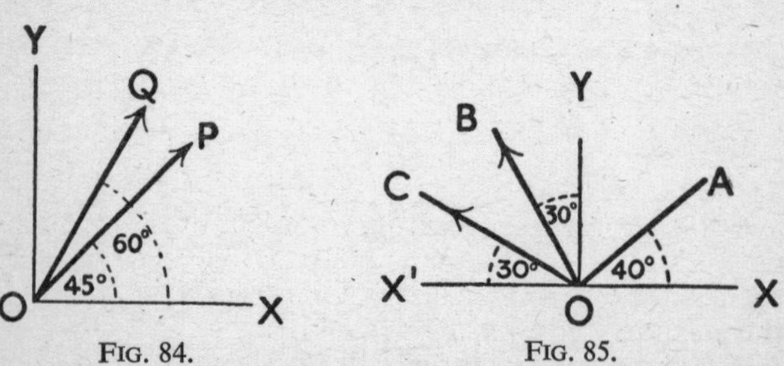

FIG. 84. FIG. 85.

2. Find the resultant of forces of 8 kgf, 4 kgf and 3 kgf acting along OA, OB and OC respectively, making angles with XOX' and OY as shown in Fig. 85.

3. Forces of 4, 5, 6 and 7 kgf act at a point O, and the angle between each successive pair is 60° (Fig. 86). Find the resultant.

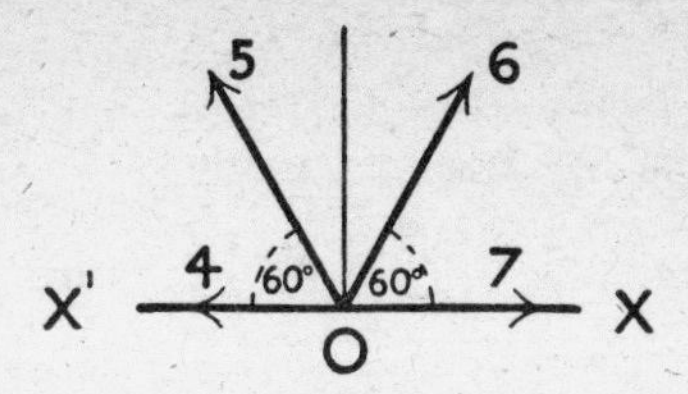

FIG. 86.

4. $ABCD$ is a square and forces of 5, 8, 10 and 6 kgf act at O, the intersection of the diagonals, and act in the direction of the corners of the square. Drawing axes of reference as shown in Fig. 87, find the resultant of the forces.

5. Three forces of 10, 4 and 5 mgf act at a point O and make angles with the straight line XOX' as shown in Fig. 88. Find the resultant.

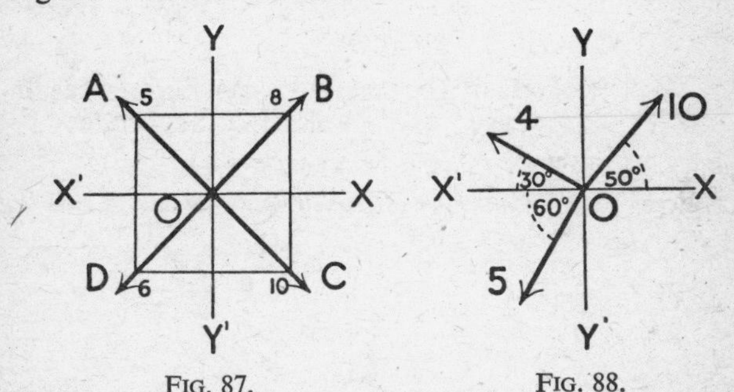

FIG. 87. FIG. 88.

6. Six strings, each making an angle of 60° with the next, are attached to a small body. If the tensions in the strings are 1, 2, 3, 4, 5 and 6 units respectively, what is the resultant force on the body?

7. From a point O straight lines OX, OA, OB, OC are drawn so that $\angle XOA = 40°$, $XOB = 120°$, $XOC = 135°$. Forces act along these lines as follows: 12 gf along OX, 10 gf along OA, 8 gf along OB and 6 gf along OC.

Find the magnitude of the resultant of these forces, and the angle made by the line of action with OX.

8. Three forces of 10, 8 and 6 tf act at a point, and the angle between each pair is 120°. Find the magnitude of the resultant, and the angle the resultant makes with the 10 tf.

Note. Remember that the resultant must be specified by magnitude and direction.

CHAPTER VI

TRIANGLE OF FORCES; POLYGON OF FORCES. LAMI'S THEOREM

68. The Triangle of Forces

In § 47 an experiment was described which had for its purpose the investigation of the theorem of the Parallelogram of Forces. That experiment, in a somewhat different form, may be used to illustrate another aspect of that theorem.

Fasten three spring-balances to hooks on a horizontal (or vertical) board. Join them by strings to a small ring so that they are all in tension. Arrange them so that they register different tensions and rest in equilibrium (Fig. 89).

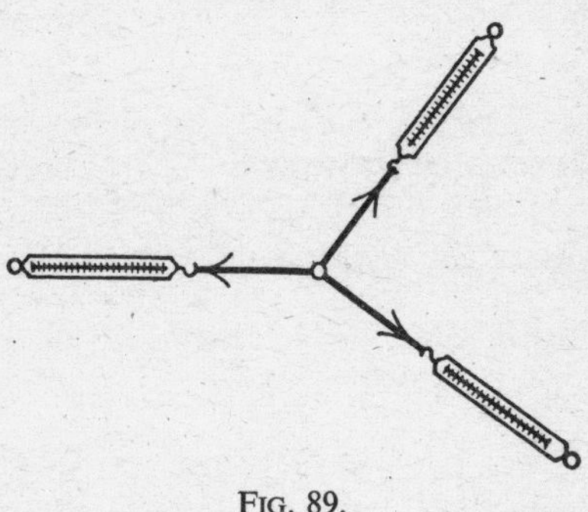

FIG. 89.

There will thus be three forces, represented by the tensions in the strings, acting on the small ring and in equilibrium.

The tensions in the strings are registered on the spring-balances.

Place a piece of paper under the strings and draw three lines on it corresponding to the strings. The three lines will, of course, be concurrent at a point corresponding to the small ring.

A diagram will thus be obtained similar to Fig. 90(*a*), in which the three straight lines which are concurrent at O represent the directions of the three forces F_1, F_2 and F_3, whose magnitudes are registered on the spring-balances.

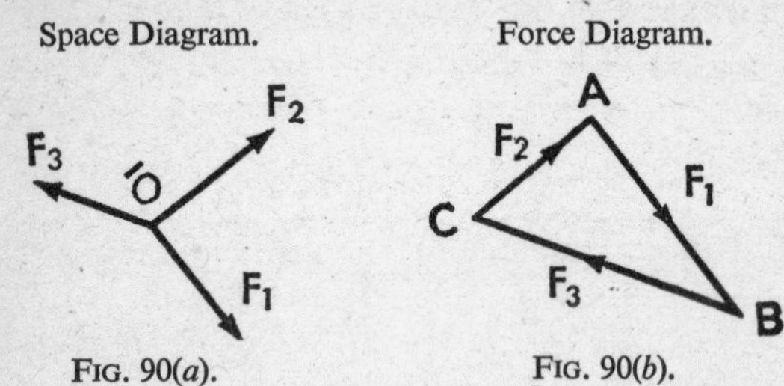

FIG. 90(*a*). FIG. 90(*b*).

Now draw a triangle ABC as in Fig. 90(*b*), having:

AB parallel to the direction of the force F_1.
BC parallel to the direction of the force F_3.
CA parallel to the direction of the force F_2.

These two diagrams are called the **Space diagram** and the **Force diagram** respectively.

Now

(1) **Measure the lengths of the sides of the triangle** ABC.

(2) **Find the ratios of the lengths of the sides to the forces to which they are parallel,**

i.e. the ratios $\frac{F_1}{AB}, \frac{F_2}{BC}, \frac{F_3}{CA}$.

It will be found that, with allowances for slight errors, these ratios are equal.

Repeat the experiment with different positions of the spring-balances and different tensions indicated by them.

It will be found that in every case the ratios obtained are equal—i.e.:

$$\frac{F_1}{AB} = \frac{F_3}{BC} = \frac{F_2}{CA}.$$

This may also be expressed thus:

$$\frac{F_1}{F_3}=\frac{AB}{BC}.$$

$$\frac{F_1}{F_2}=\frac{AB}{CA}.$$

69. These results can be expressed as follows:

The lengths of the sides of the triangle *ABC* are proportional to the magnitudes of the forces which they represent.

This can be expressed in the following form:

If the length of *AB* be taken to represent F_1, then the length of *BC*, in the same scale, will represent F_3 and the length of *CA* will represent F_2.

The forces F_1, F_2 and F_3 are in equilibrium, therefore the above results may be expressed in general form as follows, the theorem being known as the Triangle of Forces:

The Triangle of Forces

If three coplanar forces, acting at a point, are in equilibrium, and straight lines be drawn parallel to the directions of the forces, then the lengths of the sides of the triangle so formed are proportional to the magnitudes of the forces which they represent.

70. It should be noted that in Fig. 90(*b*) the directions in which the forces act, as indicated by the arrow-heads on the sides of the triangle, follow the same way round in order: all will be clockwise or all will be anti-clockwise.

If, however, the direction of one of these forces be reversed, as in 91(*b*), then this force can be regarded as the resultant.

Thus in Fig. 91(*a*) **the force *R* maintains equilibrium** in conjunction with F_1 and F_2 and is called the **equilibriant.**

In Fig. 91(*b*), the force F_3, with direction reversed, is the **resultant** of F_1 and F_2.

It corresponds to the diagonal of the parallelogram in the theorem of the Parallelogram of Forces.

Generally, when three forces acting at a point can be represented in magnitude and direction by the sides of a triangle taken in order, **any one of the forces represents either the resultant or equilibriant of the other two,** according to the direction in which it acts.

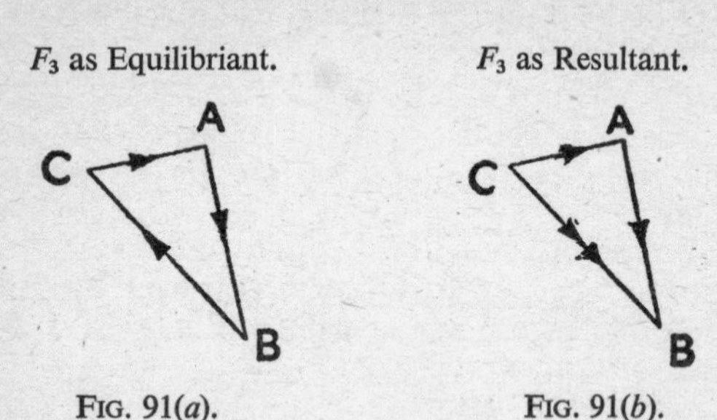

FIG. 91(*a*). FIG. 91(*b*).

It will be seen that the triangle of force corresponds to the triangle *OBC* (Fig. 56), which is part of the parallelogram of force, but the method of approach is a converse one.

In the case of the parallelogram of forces two straight lines were drawn which represented two forces in **magnitude and direction.** It was then found that the third side of a triangle, in the first instance the diagonal of the parallelogram, represented the resultant of these two forces.

But in the triangle of forces experiment it was found that, if a triangle was formed whose sides were **parallel to the directions only,** then the **lengths of these sides were proportional to the magnitudes of the forces.**

71. The inclined plane

In § 60 we considered the forces acting on a body resting in equilibrium on a smooth surface inclined to the horizontal. This is usually called an **inclined plane.** The triangle of force provides a convenient way of representing the forces which act on the body. The following two cases will serve to illustrate the method.

(1) Applied force parallel to the plane

In Fig. 92 let a body O, weight W, rest on a smooth inclined plane LMN.

Let θ denote $\angle LMN$, the angle of the slope of the plane.

Let O be held on the plane in equilibrium by a force F acting up the plane.

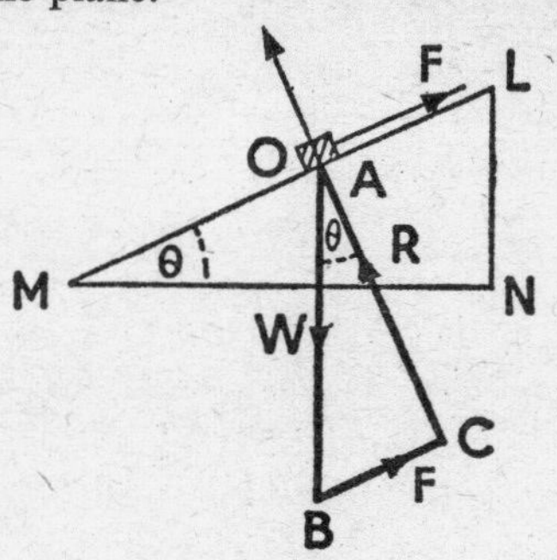

FIG. 92.

Then the forces acting on O are:

(*a*) W acting vertically downwards.
(*b*) F acting parallel to the plane.
(*c*) R, the reaction of the plane, perpendicular to the plane.

Since these are in equilibrium, they can be represented by a triangle of force.

To draw the triangle of force

Produce the line of action of W so that, on a suitable scale, AB represents the force of W kgf.

Draw AC perpendicular to LM.

From B draw BC perpendicular to AC, and therefore parallel to the direction of the force F.

Then the triangle ABC has its sides parallel to the forces acting on O and the $\angle ABC = \theta$.

$\therefore$ By the theorem of the Triangle of Forces, the sides are proportional to the magnitudes of the force W, F and R.

But AB was drawn to represent the magnitude of W.

$\therefore$ With the same scale:

AC represents R

and BC represents F.

$\therefore$ $\frac{F}{W} = \frac{BC}{AB} = \sin\theta$. $\therefore$ $\boldsymbol{F = W \sin\theta}$.

Also $\frac{R}{W} = \frac{AC}{AB} = \cos\theta$. $\therefore$ $\boldsymbol{R = W \cos\theta}$.

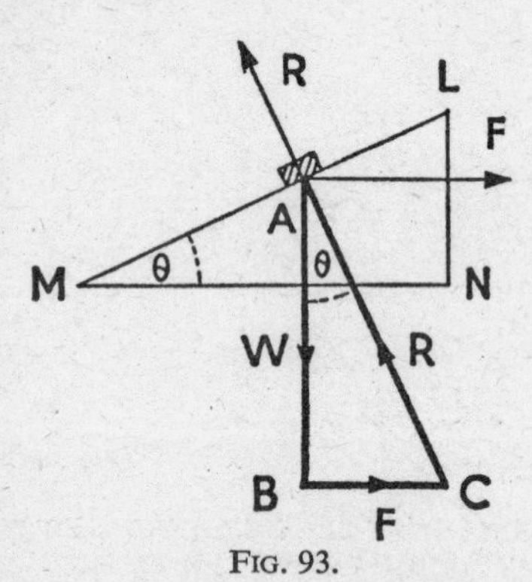

FIG. 93.

These are the resolved parts of W down the plane and perpendicular to it as shown in § 60.

(2) Applied force acting horizontally

In this case F acts horizontally and is $\therefore$ at right angles to the direction of W.

The triangle of force, ABC (Fig. 93), is drawn as before, except that BC is drawn perpendicular to AB.

Then, as in the last case:

AB represents W

AC represents R

BC represents F.

Then $\frac{F}{W} = \frac{BC}{AB} = \tan\theta$. $\therefore$ $\boldsymbol{F = W \tan\theta}$.

$\frac{R}{W} = \frac{AC}{AB} = \sec\theta$. $\therefore$ $\boldsymbol{R = W \sec\theta}$.

72. Worked example

A weight of 20 kgf suspended by two fine light wires, 5 m and 6 m in length are fastened to two points A and B, which are 8 m apart on a horizontal beam (Fig. 94(a)). The wires are knotted at C. The weight is also fastened on at C. Find the tension in the wires AC and CB.

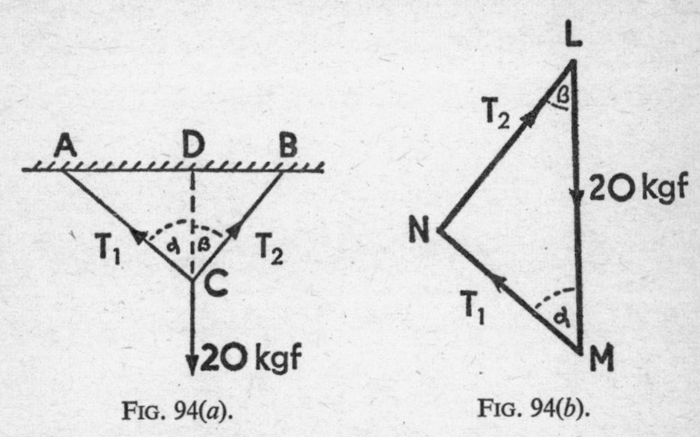

FIG. 94(*a*). FIG. 94(*b*).

Fig. 94(*a*) represents the diagram of the arrangement of the forces according to the data.

CD, the perpendicular from *C* on *AB*, is the continuation of the line of action of the weight of 20 kgf. T_1 and T_2 are the tensions in *CA* and *CB*.

(1) **Practical solution**

Construct the triangle of force for the forces T_1, T_2, and the weight of 20 kgf.

Draw *LM* to represent the 20 kgf, on a suitable scale.

From *M* draw a straight line parallel to *CA* (force T_1).

From *L* draw a straight line parallel to *BC* (force T_2).

Let *N* be the point of intersection of these straight lines.

Then ΔLMN is the triangle of force and the sides are proportional to the three forces.

$\therefore$ on the same scale that *LM* represents 20 kgf

MN will represent the tension T_1.
NL will represent the tension T_2.

(2) Trigonometrical solution

We must solve the ΔLMN, knowing that LM represents 20 kgf.

Then we can calculate the lengths of LN and MN.

From the properties of parallel lines the angles α and β as shown in the triangle of force are equal to α and β as shown in the space diagram.

To find α and β we must first solve ΔABC, knowing all the sides.

Applying the cosine rule (*Trigonometry*, § 102):

$$\cos A = \frac{8^2 + 6^2 - 5^2}{2 . 8 . 6},$$

whence $A = 38° 37'$,

and $\alpha = 90° - A = 51° 23'$.

Similarly $\cos B = \dfrac{8^2 + 5^2 - 6^2}{2 . 8 . 5}$

and $B = 48° 30'$.

$\therefore$ $\beta = 90 - B = 41° 30'$.

Also $\angle LMN = 180° - (\alpha + \beta) = 87° 7'$.

To find the sides of ΔLMN we use the sine rule (*Trigonometry*, § 90):

$$\frac{LN}{20} = \frac{\sin 51° 23'}{\sin 87° 7'},$$

whence $LN = 15{\cdot}6$ (approx.),

and $\dfrac{MN}{20} = \dfrac{\sin 41° 30'}{\sin 87° 7'},$

whence $MN = 13{\cdot}3$ (approx.).

$\therefore$ $\mathbf{T_2 = 15{\cdot}6}$ **kgf.**

$\mathbf{T_1 = 13{\cdot}3}$ **kgf.**

73. Lami's Theorem

Trigonometrical solutions of problems involving the triangle of forces are frequently made easier by using the theorem below:

Fig. 95(*a*) represents the directions of three forces F_1, F_2, F_3, in equilibrium.

Fig. 95(*b*) represents the Δ of force corresponding to these forces.

If, with the usual notation, a, b, c represent the sides of ΔABC, opposite respectively to the angles A, B, C, then by the theorem of the Triangle of Forces, the lengths of these sides are proportional to the forces F_1, F_2, F_3.

$$\therefore \qquad \frac{F_1}{a} = \frac{F_2}{b} = \frac{F_3}{c} \qquad \ldots\ldots (1)$$

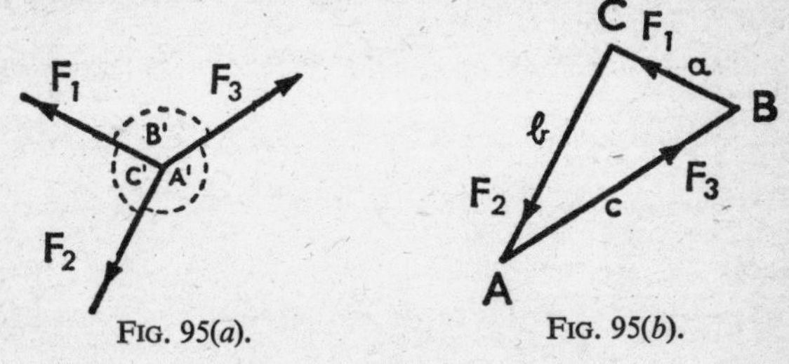

FIG. 95(a). FIG. 95(b).

Using the sine rule (*Trigonometry* § 90) for the ΔABC:

$$\frac{a}{\sin A} = \frac{b}{\sin B} = \frac{c}{\sin C}.$$

Let each of these ratios $= k$.

I.e. $$\frac{a}{\sin A} = \frac{b}{\sin B} = \frac{\sin C}{c} = k.$$

$$a = k \sin A$$
$$b = k \sin B$$
$$c = k \sin C.$$

Substituting these for a, b, c in (1) above, we have

$$\frac{F_1}{k \sin A} = \frac{F_2}{k \sin B} = \frac{F_3}{k \sin C}.$$

As these are all equal, if each be multiplied by k the results are equal.

$$\therefore \qquad \frac{F_1}{\sin A} = \frac{F_2}{\sin B} = \frac{F_3}{\sin C}.$$

Comparing with the space diagram:

$\angle A$ and $\angle A'$ are supplementary
$\angle B$ and $\angle B'$ are supplementary
$\angle C$ and $\angle C'$ are supplementary

and the sine of an angle is equal to the sine of its supplement.

$$\therefore \qquad \frac{F_1}{\sin A'} = \frac{F_2}{\sin B'} = \frac{F_3}{\sin C'},$$

where A' is the angle between F_2 and F_3,
B' is the angle between F_1 and F_3,
C' is the angle between F_1 and F_2.

This is the theorem known as:

Lami's Theorem

If three forces acting at a point are in equilibrium, each is proportional to the sine of the angle included between the other two.

This theorem is named after Bernard Lami, who enunciated it in his *Traité de Mechanique*, published in 1687.

74. Worked example

A mass of 10 *kg is supported by two strings, OA, OB, knotted at O and attached at A and B to a horizontal beam. The angle between the strings is* 120° (see *Fig.* 96(*a*)), *that*

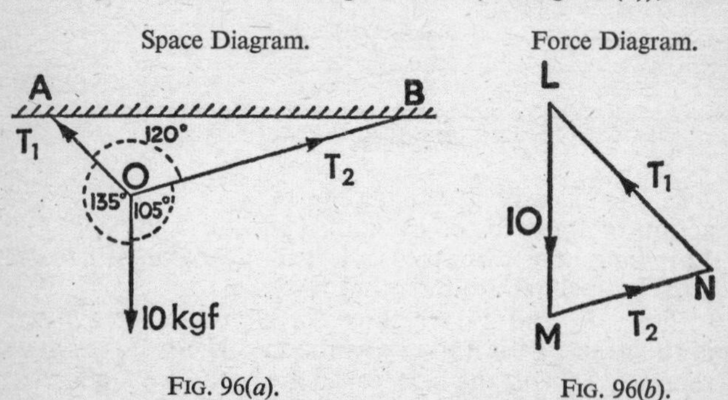

FIG. 96(*a*). FIG. 96(*b*).

between OA and the vertical string holding the weight is 135°, *and that between OB and the vertical string is* 105°. *Find the tensions in the strings OA and OB.*

Let T_1, T_2 represent the tensions in OA and OB.
Construct the triangle of force by drawing
LM parallel to vertical string and representing 10 kgf.
MN parallel to OB.
LN parallel to AO.

(1) Drawing solution

A solution can be obtained by drawing the triangle of force to a suitable scale and making LM 10 units in length to represent the weight.

The lengths of LN and MN in the scale chosen give the values of T_1 and T_2.

(2) Solution using Lami's Theorem

By this theorem:

$$\frac{T_1}{\sin 105°} = \frac{T_2}{\sin 135°} = \frac{10}{\sin 120°},$$

whence $$T_1 = \frac{10 \times \sin 105°}{\sin 120°}$$

$$= \frac{10 \sin 75°}{\sin 60°} \text{ (sin sin } \theta = \sin (180° - \theta))$$

$$= \frac{9{\cdot}659}{0{\cdot}8660}.$$

Whence, using logs

$T_1 =$ **11.15 kgf.**

Similarly $T_2 =$ **8·166 kgf.**

Exercise 9

1. The sides of a triangle are parallel to the lines of action of three forces in equilibrium, and are 8, 10 and 12 m long. The force to which the longest side is parallel is 20 kgf. What are the other forces?

2. F_1, F_2 and F_3 are three forces in equilibrium; F_1 is 10 kgf and F_2 is 15 kgf. In the triangle of force which represents them, the side parallel to the direction of F_2 is 3·75 m, and the side parallel to the direction of F_3 is 4·5 m. Find F_3 and the length of the side parallel to F_1.

3. A 2·8 kg mass is supported by two cords, one of which makes an angle of 60° with the vertical. What must

be the direction of the second cord so that its tension may be least? Find the tension in each cord in this case.

4. A metal cylinder of mass 3 kg lies on a smooth plane inclined at 35° to the horizontal, and is kept in equilibrium by a force of F kgf acting along a string which is parallel to the plane. Construct a triangle of force, corresponding to these forces, and from it find the tension in the string and the reaction at the plane.

Check by calculating the value of these forces.

5. A body weighing 5 kgf is kept at rest on an inclined plane, of angle 30°, by a horizontal string. Construct the corresponding triangle of force, and from it determine the tension in the string and the reaction at the plane.

Check by calculating the values of these forces.

6. Three forces, F_1, F_2 and F_3, act at a point and are in equilibrium. F_1 is 20 kgf. The angle between F_1 and F_2 is 105°, and that between F_1 and F_3 is 120°. Find the forces F_2 and F_3.

7. Two strings, AB and AC, are fastened to a horizontal beam at B and C. They are knotted at A to a third string, which hangs vertically and sustains a weight of 10 kgf. AB and AC make angles of 50° and 60° on either side of the vertical through A. Find the tensions in AB and AC.

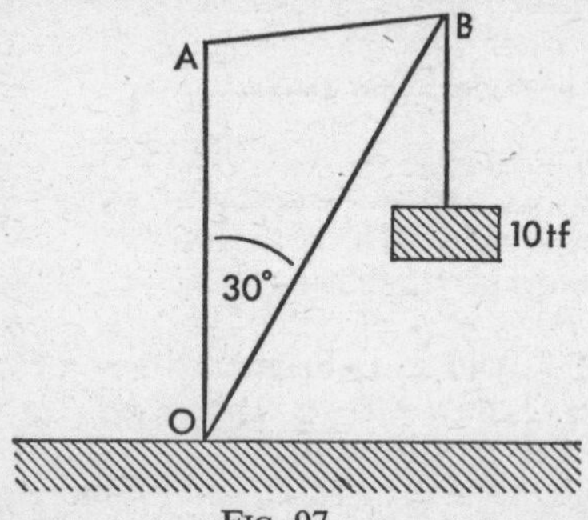

FIG. 97.

8. OB (Fig. 97) represents the jib of a crane, and from B is suspended a load of 10 tf. OB makes an angle of 30° with the vertical and AB is a tie rod. OA is 10 m long and OB is 12 m. Find the tensions in OA and AB, either by the use of Lami's Theorem or by a graphical method.

75. The Polygon of Forces

We have seen that if three forces acting at a point are in equilibrium they can be represented by the sides of a triangle in accordance with the Triangle of Forces. This principle can be extended to any number of forces.

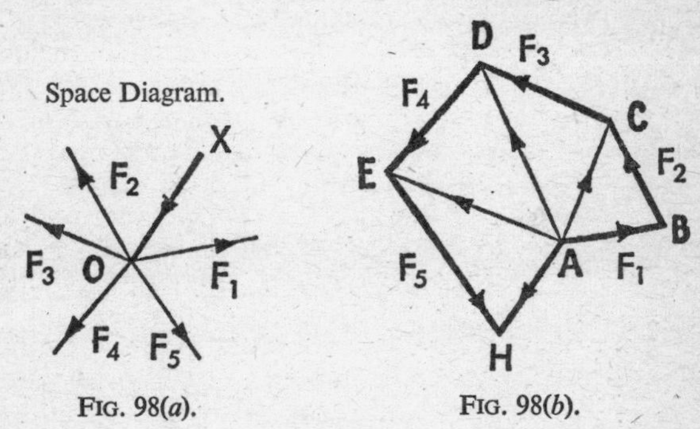

FIG. 98(*a*). FIG. 98(*b*).

Let forces F_1, F_2, F_3, F_4 and F_5, whose magnitudes are known and which act at a point O (Fig. 98(*a*)), be in equilibrium, their directions being indicated in the figure by the straight lines as marked.

(1) Taking any point A (Fig. 98(*b*)), draw AB to represent F_1 in magnitude and parallel to its direction.

(2) From B draw BC to represent F_2 in magnitude and parallel to its direction.

Then ABC is a Δ of force, and AC **represents the resultant of F_1 and F_2 in magnitude and direction.**

(3) From C draw CD to represent F_3 in magnitude and parallel to its direction.

Then ACD is a Δ of force for F_3 and AC which is the resultant of F_1 and F_2; AD is $\therefore$ the resultant of AC and CD.

Then AD represents the resultant of $F_1 + F_2 + F_3$ in magnitude and direction.

(4) From D draw DE to represent F_4 in magnitude and direction.

Then, as in the other cases:

***AE* represents the resultant of $F_1 + F_2 + F_3 + F_4$ in magnitude and direction.**

(5) From *E* draw *EH* to represent F_5 in magnitude and direction.

Then *AH* represents the resultant of *AE* and *EH*.

∴ *AH* represents the resultant of $F_1 + F_2 + F_3 + F_4 + F_5$ in magnitude and direction.

I.e. *AH* represents in magnitude and is parallel to the direction of the resultant of the system of forces acting at *O*.

∴ if from *O* in Fig. 98(*a*) we draw *OX* parallel to *AH* and of the same length it will represent the resultant of the forces concurrent at *O*.

Since *AH* represents the resultant of the five forces, then the force which it represents in magnitude, with the **direction reversed**—i.e. *HA*—if applied at *O* would represent the **equilibriant** of the five forces, and the system will be in equilibrium.

This is the theorem known as the Polygon of Forces, and it may be defined as follows:

Polygon of Force

If a number of forces acting at a point are in equilibrium, they can be represented in magnitude and direction by the sides of a polygon, taken in order.

For example, **four** forces acting at a point and in equilibrium can be represented by the sides of a **quadrilateral.**

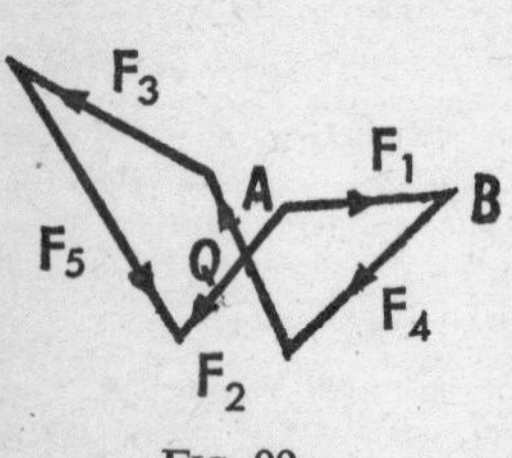

Fig. 99.

It should be noted that the words "taken in order" in the definition above mean, as in the case of the triangle of force, that the direction of the forces going round the polygon is the same, all the arrow-heads pointing clockwise or anti-clockwise.

If this condition be fulfilled, the forces can be drawn in any order, so that the same polygon of force may have different shapes.

Thus in the example above the polygon might be

drawn as in Fig. 99. Starting from A, the sequence of forces is F_1, F_4, F_2, F_3, F_5, all of which follow the same order round. Then, as in all cases:

The force which closes the polygon is the Resultant or Equilibriant according to the sense in which it acts.

In working problems the unknown force should be drawn last. Then it closes the polygon and its direction and magnitude are determined.

76. Worked example

Find by means of a polygon of force the resultant of four forces of 4 *kgf,* 3 *kgf,* 3 *kgf and* 4 *kgf, acting as shown in Fig.* 100.

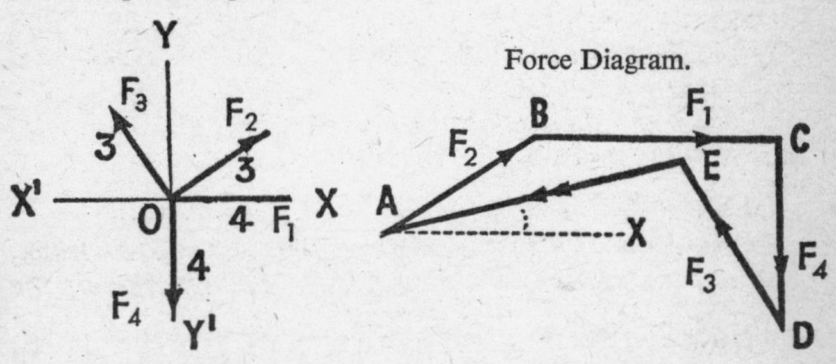

FIG. 100(*a*). FIG. 100(*b*).

To draw the Polygon of Force

From a suitable point A draw AB, parallel to F_2 and equal to 3 units of length.

Then in succession draw:

BC parallel to F_1,
CD parallel to F_4,
DE parallel to F_3.

All must be drawn to the scale adopted for Q.

Join EA.

Then EA **represents the resultant.**

Its length is 5.13.

∴ Resultant is **5·13 kgf.**

The angle made with OX is given by the angle EAX in the force diagram. It is about 12°.

The result may be checked using the method of § 66.

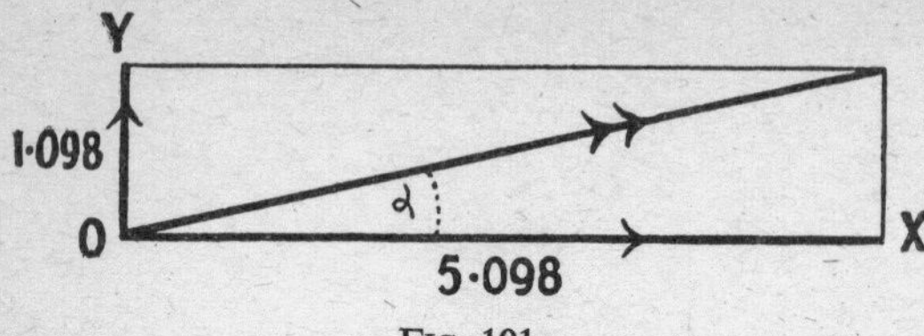

FIG. 101.

Taking the resolved parts we get a total of

5·098 kgf along OX,
1·098 kgf along OY.
(*See* Fig. 101.)

$$\therefore \text{ Resultant} = \sqrt{5{\cdot}098^2 + 1{\cdot}098^2}$$
$$= 5{\cdot}13 \text{ approx.}$$

And
$$\tan \alpha = \frac{1{\cdot}098}{5{\cdot}098} = \tan 12^\circ\, 8'.$$
$$\therefore \qquad \alpha = 12^\circ\, 8'.$$

Exercise 10

1. Forces represented by $OA = 8$ kgf, $OB = 10$ kgf and $OC = 5$ kgf all act at O and are separated by angles

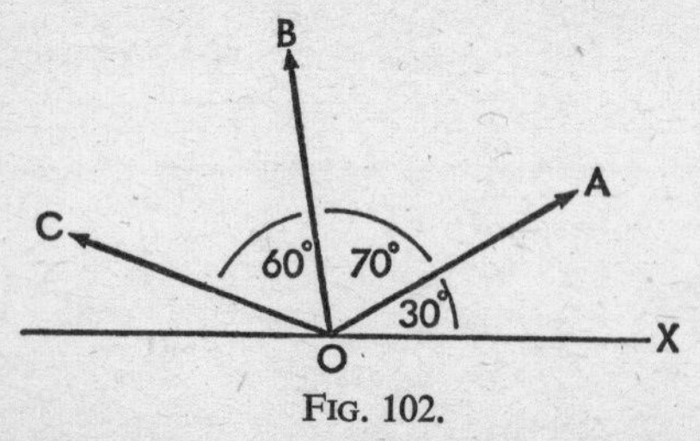

FIG. 102.

as shown in Fig. 102. Find graphically the magnitude of the resultant and the direction which it makes with OX.

2. Three forces, $F_1 = 30$ kgf, $F_2 = 40$ kgf and $F_3 = 50$ kgf act at a point. The angle between F_1 and F_2 is

45°, and that between F_2 and F_3 is 60°. Find the magnitude of the resultant, graphically and by calculation.

3. Three forces pass through a point in a body and act outwards. The forces are F of 5 kgf, F' of 10 kgf, acting

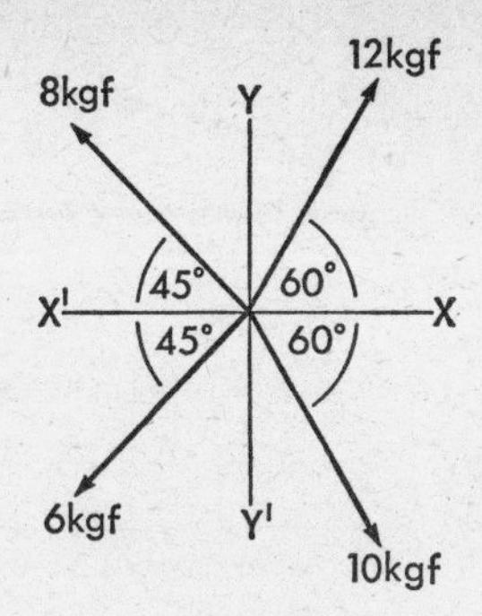

FIG. 103.

at 60° to F, and F'' of 15 kgf at 210° to F measured in the same direction. Is the body in equilibrium? If not, what is the resultant of the forces, and what single force will produce equilibrium?

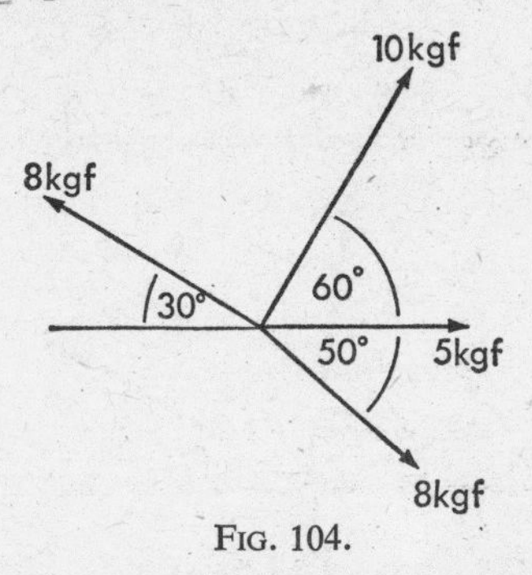

FIG. 104.

4. Four forces act at a point O as shown in Fig. 103. Find the resultant and the angle which it makes with OX.

5. Forces of 8, 5, 10 and 8 kgf act at a point and in directions as shown in Fig. 104. Find the magnitude of

the resultant and the direction it makes with the 5 kgf force.

6. Forces of 2 kgf and 4 kgf act along *OA* and *OB*, and the angle between them is 120°. A third force acts along *OC* which makes angles of 120° with both *OA* and *OB*. The direction of the resultant is perpendicular to *OB*. What is the magnitude of the force along *OC*?

7. Show that if three equal forces acting at a point are in equilibrium the angle between the lines of action of any two is 120°.

CHAPTER VII

FRICTION

77. Friction as a force

When considering problems connected with forces which act on bodies moving or lying on surfaces, or cords passing over pulleys, we have assumed that the surfaces were smooth. By this assumption the problems were not complicated by the forces brought into play when surfaces are not smooth.

It is common experience that all surfaces are rough to a greater or less degree, and that, in consequence, there is resistance to motion over them.

If a body is resting on a horizontal plane, the only forces acting on it to keep it in equilibrium are:

(1) Its weight acting vertically downwards.
(2) The reaction of the surface, equal and opposite to the weight.

Neither of these forces can have any resolved part at right angles to itself, and consequently has no effect in a horizontal direction.

The resistance to horizontal motion, which we know from experience is brought into play, must be due to the force, which is called the **force of friction.**

This force may be very useful, even essential, or it may be the reverse. Without it, walking and running would be very difficult, as we know if we try to move on a surface which is smooth and offers very little resistance, such as ice. But for the parts of moving machinery we employ metals with surfaces as smooth as possible to minimise friction.

Without friction the wheels of a locomotive or of a motor would spin round rapidly, without any progress being made; but the hull of a fast liner or the skin of an aeroplane are made as smooth as possible, so as to reduce frictional resistance to a minimum.

It is necessary, then, to investigate the action of the resistance due to friction, and to know how to measure it.

78. Limiting friction

For the measurement of this force we must rely on experiments.

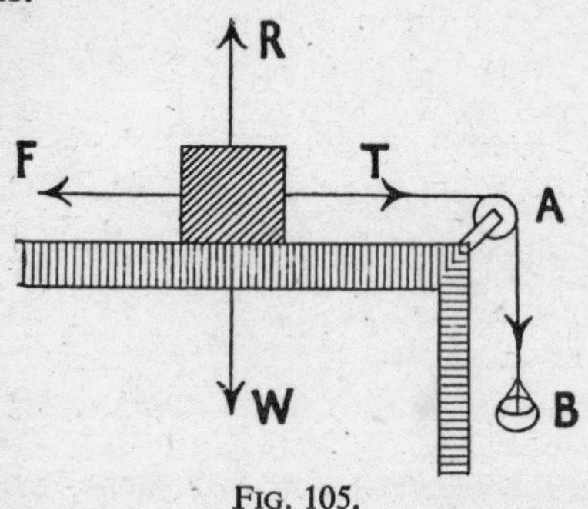

FIG. 105.

Experiment. A block of wood of known weight, W kgf, is placed on a wooden table (Fig. 105). As it lies on the table, the forces acting on it, as stated above, are:

(1) Its **weight,** W**,** acting vertically downwards.

(2) The **reaction** of the table, R, equal and opposite to W.

A light cord is attached to the block, and carried over a small smooth pulley, A, to a scale-pan B, whose weight is known.

A small weight is placed in the scale-pan. This introduces a new force, which, together with the weight of the scale-pan, acts vertically downwards. It thus sets up a tension in the string which is transmitted, practically unaltered, and acts horizontally on the block.

If the weight be small, the block will not move.

Consequently a force must have been brought into action—the force of friction—which acts horizontally, and, since there is no motion, is equal and opposite to T. Thus we conclude that the force of friction begins to act only when there is a pull in the cord (T), and ceases when the pull ceases.

We now add more weights gradually to the scale-pan: T is thus increased, and so long as there is no motion the force of friction is also increased, so as to equal T.

As we continue to increase the weights in the pan, a time will come when the block will just begin to move. At that moment the two forces are equal, but with a slight increase in T the block moves.

Thus the force of friction will increase up to a certain amount, and no more.

There is therefore a **limiting value of the force of friction** beyond which it cannot increase, and the body will not move until the force applied to move the body (T) is greater than it.

79. Coefficient of friction

Let F be the limiting value of friction in the above experiment.

Let N be the normal reaction of the board on the block. Then the forces acting on the block in the experiment are (*see* Fig. 106):

(1) **Vertically**:

W, the **weight** downwards.
N, the **reaction** upwards.

(2) **Horizontally**:

T, the pull of the scale-pan.
F, force of friction opposite in direction to T.

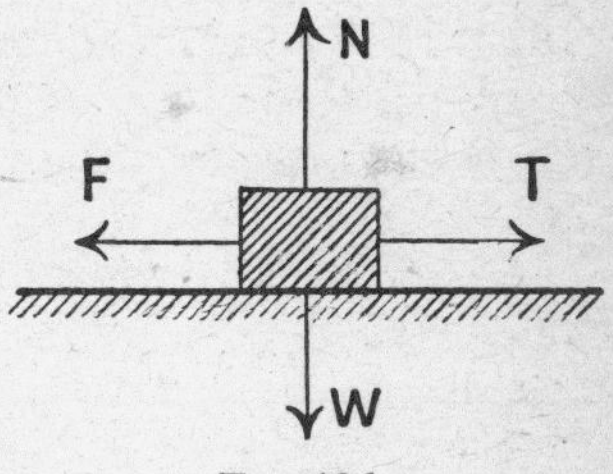

FIG. 106.

As there is equilibrium

$$F = T.$$

Continuing the experiment place **additional weights on the top of the block.**

Then it will be found that more weights must be placed on the pan until the point is reached when the block is again about to move:

i.e. as **W increases,**
F also increases.

As further weights are added, both increase; but in every case it will be found that, with slight errors, the ratio:

$$\frac{F}{W} \textbf{ is constant.}$$

Also $W = N$ throughout the experiments.

$$\therefore \qquad \frac{F}{N} \text{ is constant.}$$

$\therefore$ as F increases, **the ratio of F to the normal reaction N remains constant.**

This constant value of the ratio applies only to the materials used in the experiment.

If two metal surfaces were used it would be different.

The value of $\frac{F}{N}$ for a given pair of surfaces is usually denoted by the Greek letter μ (pronounced "mew").

We can therefore write:

$$\frac{F}{N} = \mu,$$

or

$$F = \mu N.$$

In another form this means:

The limiting value of the force of friction for any two substances is equal to the total normal reaction multiplied by μ.

The constant μ is called the **coefficient of friction.**

Values of μ.

The following are values of μ for a few substances.

Wood on wood 0·35 (if polished) to 0·5 (if dry).
Wood on metal 0·10 (if polished) to 0·6.
Metal on metal 0·15 to 0·18.
Metal on metal (if greased) 0·1.
Wood on stone 0·6.
Leather on wood 0·62.

As indicated above, the constant μ may vary for two substances, according to the degree of polish of the surfaces.

It **does not depend on the areas of the surfaces in contact.**

A smooth surface is one whose coefficient of friction is zero or is so small as to be negligible.

80. The angle of friction

The results of the last paragraph showed the relation between:

(1) the limiting friction, F, and
(2) the normal reaction,

viz. $$\frac{F}{N} = \mu.$$

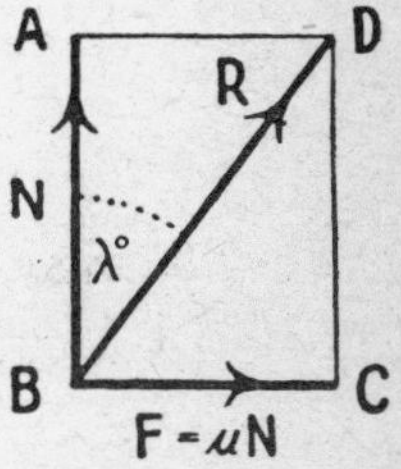

The resultant of these two forces F and N can be found by means of the Parallelogram of Forces. As the two forces are at right angles, the parallelogram will be a rectangle.

FIG. 107.

In Fig. 107:

AB represents N,
BC represents $F = \mu N$.
$\therefore$ BD represents R, the resultant.

Let the angle between N and R be denoted by λ (pronounced "lambda").

Then $$\tan \lambda = \frac{F}{N} = \frac{\mu N}{N} = \mu.$$

The angle λ is called the ***angle of friction.***

Thus, *the angle of friction is the angle whose tangent is equal to the coefficient of friction*

or $$\boldsymbol{\lambda = \tan^{-1} \mu.}$$

81. Kinetic friction

When motion takes place, the friction, in general, is less than the limiting friction, which tends to prevent motion. This is called "kinetic friction", and it depends on the velocity at which the body moves.

82. The laws of friction

From the conclusions reached above we may formulate the laws of friction as follows:

I. Friction acts in an opposite direction to that in which a body tends to move.

II. The force of friction varies, while there is equilibrium, according to the magnitude of the applied force, but cannot exceed that which is just sufficient to prevent the body from moving.

III. The limiting friction depends on the nature of the surfaces in contact, but is constant for two given materials.

This is equal to the ratio

$$\frac{\textbf{total horizontal reaction}}{\textbf{total normal reaction}}$$

and is denoted by μ.

IV. If a body rests on a surface the *resultant* reaction of the surface on the body is inclined to the vertical at an angle λ, such that $\tan \lambda = \mu$.

83. Action on a body about to move on a horizontal surface by a force inclined to the surface

Let a wooden block, weight W, resting on a horizontal surface (Fig. 108) be acted upon by a force F, inclined to horizontal at angle θ.

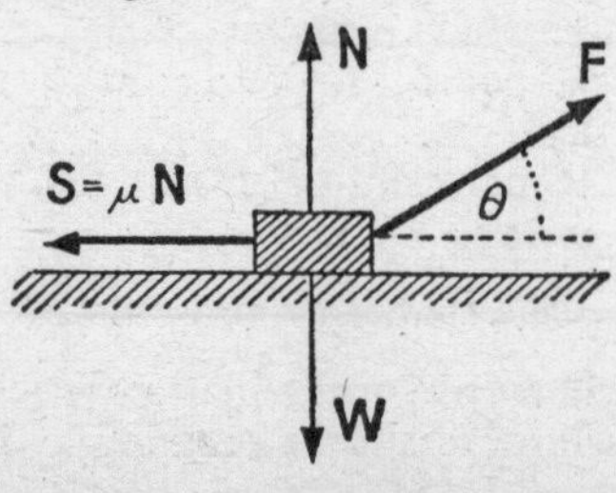

FIG. 108.

Let N be the normal reaction.

Then $$N = W.$$

Let S be the force of friction acting between the surfaces horizontally when the body is on the point of moving.

Then $$S = \mu N.$$

Let R be the resultant of N and S.

Then R is inclined at angle λ to the vertical (§ 80), so that

$$\tan \lambda = \mu.$$

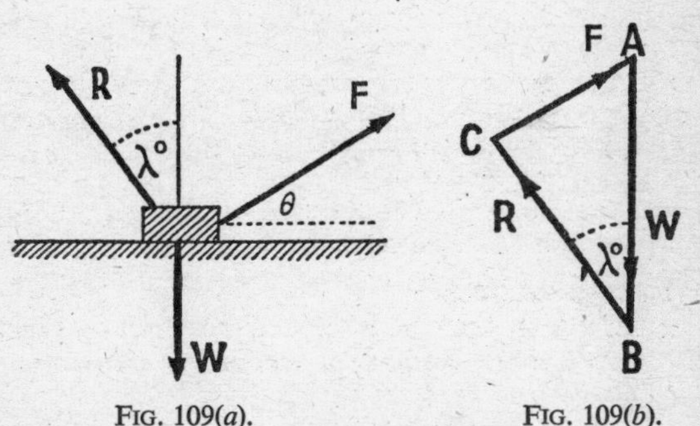

FIG. 109(a). FIG. 109(b).

Replacing N and S by their resultant, the forces acting on the block are as shown in Fig. 109(a).

The triangle of force for these forces is shown in Fig. 109(b) and is drawn as follows:

(1) Draw AB to represent the **magnitude** and **direction** of W, which is **known.**

(2) Draw BC parallel to R—i.e. making an angle $\lambda°$ with the vertical.

(3) Draw AC parallel to F, making $90° - \theta$ with the vertical and **meeting** BC **in** C.

Then on the scale that AB represents W:

AC will represent the magnitude of F,

and BC will represent the magnitude of R.

A **graphical solution** can thus be obtained.

Trigonometrical solution

To determine F, the triangle ABC must be solved.

Now $\angle ACB = 180° - (\angle BAC + \angle ABC)$
$= 180° - (90° - \theta + \lambda)$,
but $\sin \alpha = \sin (180° - \alpha)$. (*Trigonometry*, § 70.)
$\therefore \quad \sin ACB = \sin (90° - \theta + \lambda)$.

Since θ and λ are known, we can thus find ACB. Using the sine rule:

$$\frac{F}{\sin \lambda} = \frac{W}{\sin ACB}.$$

$$\therefore \quad \frac{F}{\sin \lambda} = \frac{W}{\sin (90° - \theta + \lambda)},$$

whence **F can be determined.**

Least value of F. Since F is represented by AC in the triangle of force, F will have its least value when AC is least—i.e. when **AC is perpendicular to BC.** Then the triangle of force is as shown in Fig. 110.

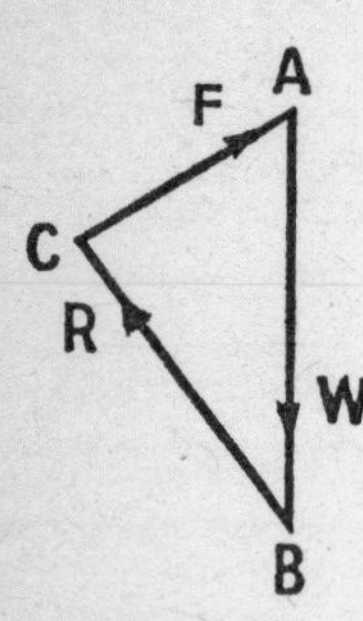

FIG. 110.

Then $\dfrac{F}{W} = \sin \lambda$.

$\therefore \quad F = W \sin \lambda$.

$\therefore$ **The least value of F is $W \sin \lambda$, when the angle which it makes with the horizontal is λ.**

84. Worked example

A metal block weighing 20 kgf rests on a horizontal board, and the coefficient of friction between the surfaces is 0·22.

Find:

(1) *The horizontal force which will just move the block.*

(2) *The force acting at* 30° *with the horizontal which will just move the block.*

(3) *The least value of the force inclined to the horizontal which will just move the block.*

(1) The **horizontal force** F_1 which will just move the block, as shown in § 79, is given by:

$$F_1 = \mu N$$

and $$N = W.$$

$\therefore$ $$F_1 = \mu W = 0{\cdot}22 \times 20$$
$$= \mathbf{4{\cdot}4\ kgf.}$$

(2) When F_2 makes 30° with the horizontal the forces acting are as shown in Fig. 111(*a*) (*see* § 83).

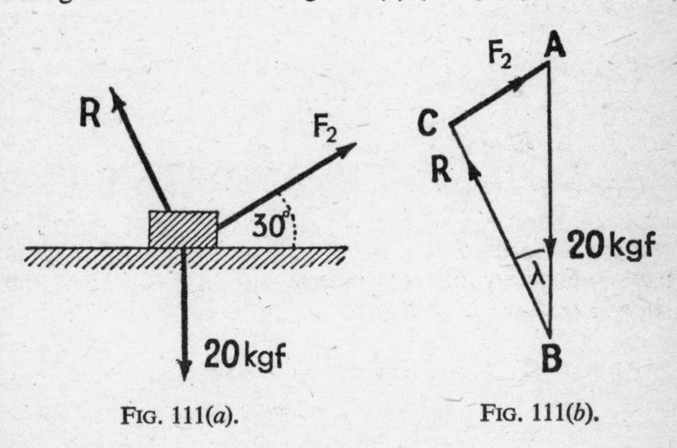

FIG. 111(*a*). FIG. 111(*b*).

To find λ:

$$\tan \lambda = \mu$$
$$= 0{\cdot}22.$$

$\therefore$ $$\lambda = \mathbf{12^\circ\ 24'}.$$

To draw the triangle of force (Fig. 111(*b*)):

(1) Draw AB vertical to represent 20.
(2) From B draw BC parallel to R—i.e. $\angle B =$ 12° 24′.
(3) From A draw AC parallel to F_2—i.e. making 30° with the horizontal and meeting BC in C.

Then AC represents F_2 and can be found from a drawing to scale.

Trigonometrical solution

The ΔABC must be solved as shown in § 83.

$$\angle ACB = 180° - (60° + 12° 24')$$
$$= 180° - 72° 24'.$$

$\therefore \sin ACB = \sin 72° 24'.$ (Supplementary angles.)

Using the sine rule:

$$\frac{F_2}{\sin 12° 24'} = \frac{20}{\sin 72° 24'}.$$

$$\therefore \quad F_2 = \frac{20 \times \sin 12° 24'}{\sin 72° 24'}.$$

$$\therefore \log F_2 = \log 20 + \log \sin 12° 24' - \log \sin 72° 24'$$
$$= \log 0{\cdot}6537.$$

$\therefore \quad F_2 = 4{\cdot}505$

or $\quad F_2 =$ **4.5 kgf nearly.**

No.	log
20	1·3010
sin 12° 24′	$\bar{1}$·3319
	0·6329
sin 72° 24′	$\bar{1}$·9792
4·505	0·6537

(3) **Least value of F_2.**

As shown in § 82, the least value of F_2 will occur when in the triangle of force AC is perpendicular to BC.

Then $\quad F_2 = 20 \sin \lambda$

$$= 20 \times \sin 12° 24'$$
$$= 20 \times 0{\cdot}2147.$$

$\therefore \quad F_2 =$ **4.29 kgf approx.**

Exercise 11

1. A block of wood of mass 6 kg rests on a horizontal table. A horizontal force of 2·5 kgf is just sufficient to cause it to slide.

Find:

(*a*) the coefficient of friction for the two surfaces;
(*b*) the angle of friction.

2. A block of wood of mass 15 kg rests on a horizontal table and can just be moved along by a horizontal force of 4·2 kgf. Another 6 kg is placed on the block. What is

the least horizontal force which will just move the new mass?

3. A body of mass 12 kg rests on a horizontal table and the coefficient of friction between the two surfaces is 0·32. What horizontal force will just move the block?

4. A 5-kg block of wood rests on a horizontal surface and the angle of friction at the two surfaces is 14°. What is the least force that will cause the block to move along the surface?

5. A block of wood resting on a horizontal table is just moved by a horizontal force of 15 kgf. If the coefficient of friction between the table and block is 0·35, find the mass of the block.

6. A block of wood of mass 5 kg rests on a rough horizontal board and the coefficient of friction between the two surfaces is 0·4. By means of a string inclined at 30° to the board, a pull is exerted on the block, and this is just sufficient to make the block move. Calculate the force exerted by the string.

7. A block of mass 40 kg is pulled very slowly along a horizontal plane, the coefficient of friction being 0·2. Find the magnitude of the pull if its line of action is (1) horizontal, (2) at 45° to the horizontal and (3) such that the pull is minimum.

8. A uniform ladder rests at an angle of 45° against a smooth vertical wall. Show that the coefficient of friction between the foot of the ladder and the (horizontal) ground must be at least 0·5.

9. A block of wood of mass 20 kg rests on a horizontal table, and the coefficient of friction between the surfaces is 0·5. Find the force, acting at 45° with the horizontal, which will just move the block.

85. The inclined plane and friction

The inclined plane provides an easy way of observing the laws of friction. When the forces acting on a body lying on an inclined plane were considered (§ 71) it was assumed that the plane was smooth. We must now see how these forces are affected when friction is taken into account.

Experiment. Take a board and place on it a wooden block, such as was used in the experiment described in § 78.

Tilt the board slowly through a small angle θ.

The weight (W) of the block can now be resolved into components:

W **sin θ** acting along and down the plane.

W **cos θ** acting perpendicular to the plane.

If the angle of tilting (θ) is small the block will not slide down the plane, but will remain at rest.

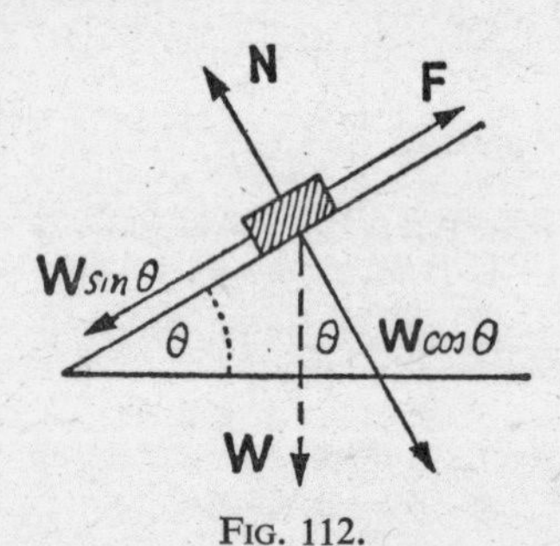

FIG. 112.

∴ The component of the weight down the plane, $W \sin \theta$, must be counteracted by a force acting along and up the plane.

This must be the **force of friction,** denoted by F.

Now sin θ increases as θ increases.

∴ As the board is tilted a little more, $W \sin \theta$ will increase, and consequently the counterbalancing force of friction, F, must increase.

But, as in the case of a body resting on a horizontal surface, this force of friction has a limiting value, beyond which it cannot increase.

∴ As the angle of the plane is increased, $W \sin \theta$ increases; F will also increase until it reaches its limiting value. Beyond that, $W \sin \theta$ becomes greater than F and the block begins to move **down** the plane.

At the point of slipping

$$F = W \sin \theta \quad . \; . \; . \; . \; . \; . \quad (1)$$

Throughout the tilting of the plane, N, **the normal reaction of the plane, decreases,** since it is equal to $W \cos \theta$, and **cos θ decreases as θ increases.**

In the limiting case, when the block is on the point of slipping

$$N = W \cos \theta \quad . \quad . \quad . \quad . \quad . \quad (2)$$

From equations (1) and (2):

$$\frac{F}{N} = \frac{W \sin \theta}{W \cos \theta} = \tan \theta.$$

$$\therefore \qquad \frac{F}{N} = \tan \theta.$$

In the limit **θ becomes the angle of friction for the two materials** and may be replaced by λ.

$$\therefore \qquad \frac{F}{N} = \tan \lambda$$

$$= \mu.$$

$$\therefore \qquad \boldsymbol{F = \mu N.}$$

In this way the angle of friction, and therefore μ, is easily found by direct observation of the angle of the plane.

Note. When solving problems connected with the inclined plane and involving the force of friction, the student should note the following two points:

(1) The **force of friction always opposes motion.** Consequently, if a body is on the point of slipping down the plane, the force of friction acts up the plane and vice versa.

(2) If a force acting on a body on an inclined plane is inclined to the plane, then it has resolved parts along and perpendicular to the plane. The latter produces a reaction of the plane and, as the force of friction applies to the **total normal reaction** of the plane, it applies to the resolved part of the force which is perpendicular to the plane.

Worked example

A body weighing 10 *kgf is on the point of slipping down a plane which is inclined at* 20° *to the horizontal. What force parallel to the plane will just move it up the plane?*

As the body is just on the point of slipping **down** the plane, the angle of the plane is equal to the angle of friction,

i.e. $\lambda = 20°$,

and $\mu = \tan\lambda = \tan 20°$.

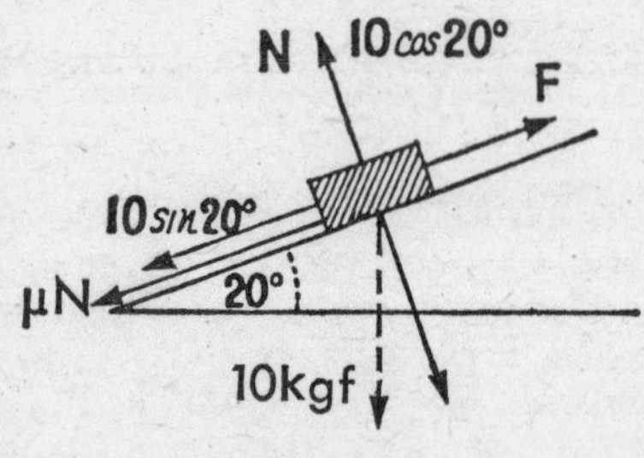

FIG. 113.

When the body is on the point of moving **up** the plane, the **forces opposing motion,** i.e. acting **down** the plane, are:

(1) Resolved part of weight,

$$W \sin\theta = 10 \sin 20°.$$

(2) Limiting value of friction, which is μN,

where $N = W\cos\theta = 10\cos 20°$.

$\therefore$ $\mu N = \mu \times 10 \cos 20°$.

The **force necessary just to move the body up the plane** must be equal to the **sum of those acting down the plane.**

$$\therefore \quad F = W\sin\theta + \mu W\cos\theta$$
$$= 10\sin 20° + \mu \times 10\cos 20°,$$

but $\mu = \tan 20°$.

$$\therefore \quad F = 10 \sin 20° + 10 \tan 20° \times \cos 20°$$
$$= 10 \sin 20° + 10 \times \frac{\sin 20°}{\cos 20°} \times \cos 20°$$
$$= 10 \sin 20° + 10 \sin 20°$$
$$= 20 \sin 20° = 20 \times 0{\cdot}3420$$
$$= \mathbf{6{\cdot}84\ kgf.}$$

Exercise 12

1. A body rests on a rough horizontal board. This is gradually tilted until, when it is inclined at 22° to the horizontal, the body begins to move down the plane.

(1) What is the coefficient of friction between the body and the plane?
(2) If the body is 5 kg, what is the magnitude of the frictional force when the body begins to slip?

2. A block of wood rests on an inclined plane, and the coefficient of friction between it and the plane is known to be 0·31. At what angle must the plane be inclined to the horizontal so that the block begins to move down the plane?

3. A mass of 50 kg rests on a rough slope, inclined at 20° to the horizontal. The least force directed up the plane which will move the mass is 56 kgf. Find:

(1) The angle of friction.
(2) The least force directed down the plane which will move the mass.

4. A 5-kg body rests on a plane inclined at 40° to the horizontal. It is kept from slipping down the plane by a force of 1·5 kgf parallel to the plane. What is the coefficient of friction?

5. A body of mass 40 kg is lying on an inclined plane of slope 30°, and it is on the point of slipping down. What force applied parallel to the plane will cause the block to begin moving up the plane?

6. A load of 50 kg resting on a rough inclined plane begins to slip when the plane is tilted to an angle of 30°

with the horizontal. What force up the plane will be necessary to keep the load from slipping down the plane if the angle of slope is increased to 45°?

7. In the last question, what force would be required to make the body begin moving up the plane?

8. What horizontal force must be applied to the load of question 6 so that it will just begin to move up the plane?

CHAPTER VIII

BODIES IN MOTION; VELOCITY

86. In the preceding chapters we have confined our attention to the consideration of bodies at rest under the action of force. We now proceed to the study of the "Dynamics" section of the subject—that is, to the consideration of bodies in motion and of the laws relating to motion.

Motion of a body means that its position is changed; it is displaced. This takes place under the action of force and, as will be learnt later, in the **direction** of the line of action of the force. Thus, bodies falling under the action of the force of gravity tend to fall vertically downwards. Displacement therefore involves magnitude and direction. It is thus a vector quantity (*see* § 45).

87. Speed and velocity

When a body is in motion the **rate at which it is moving on its path is called its speed or velocity.**

In everyday language these two terms are usually accepted as meaning the same thing, and by some writers on mechanics they are treated as alternatives. A distinction, however, is usually made between them. While **speed** is defined as the **rate** only at which a body is moving, the term **velocity** implies **direction as well as rate.**

Speed is certainly frequently used in a sense which has no reference whatever to direction. Thus an aeroplane may be described as capable of a speed of "600 km per hour", whereas the term velocity is more appropriately used in such a statement as "the wind was blowing with a velocity of 70 km per hour in a direction N.N.E.".

This is a clear and logical distinction and its adoption means that

speed is a scalar quantity, involving **magnitude only,** whereas
velocity is a vector quantity, involving **direction as well as magnitude.**

88. Units in the measurement of speed and velocity

If speed is the rate at which a body is moving, this rate must involve:

(1) The **distance** it moves.
(2) The **time** taken.

Therefore the units employed in measuring speed or velocity must be in terms of the units of distance and the units of time.

Speed may be expressed either in large units such as kilometres and hours, or in small units such as cm or m and seconds.

Thus we may speak of a speed of 400 kilometres per hour, usually written as 400 km/h, or of a speed of 20 metres per second, written as 20 m/s.

Other units of distance may be similarly employed.

For **nautical purposes** the unit employed is the **knot,** which includes both distance and time, for

$$1 \text{ knot} = 1 \text{ nautical mile per hour.}$$

The nautical mile is related to the older imperial units.

$$1 \text{ nautical mile} = 6080 \text{ ft};$$
$$1 \text{ ft} = 0{\cdot}3048 \text{ m};$$
$$\therefore \quad 1 \text{ kn} = \frac{6080 \times 0{\cdot}3048}{1000} \text{ km/h}$$
$$= 1{\cdot}85 \text{ km/h}.$$

We frequently require to change rates expressed in one set of units into rates expressed in other units, in particular km/h into m/s.

$$1\ \text{km/h} = \frac{1000}{60 \times 60}\ \text{m/s},$$

i.e. $$\mathbf{1\ km/h = \frac{10}{36}\ m/s,}$$

conversely $$\mathbf{1\ m/s = 3{\cdot}6\ km/h.}$$

89. Uniform velocity or speed

Velocity or speed may be uniform or variable.

Uniform velocity:

A body moves with uniform velocity when throughout the motion equal distances are passed over in equal times, however small or however large.

Uniform velocity is, however, a theoretical conception. A French mathematician declared that "no conception is more simple than uniform velocity; nothing is more impossible to carry out practically". Not only does the existence of such forces as gravity and friction make it very difficult to prevent variations, but it is practically impossible to measure either time or distance with absolute accuracy; consequently we cannot attain to absolute accuracy in experimental results.

However, the conception of uniform velocity is a very useful and important one, and is constantly employed.

The **rate of velocity** which is uniform is obtained by **dividing any distance travelled by the time taken.**

Thus $$\textbf{velocity} = \frac{\textbf{distance moved}}{\textbf{time taken}}.$$

Thus if a body moves 28 m in 3·5 s

$$\text{velocity} = \frac{28\ \text{m}}{3{\cdot}5\ \text{s}} = 8\ \text{m/s}.$$

This may be generalised.

Let $s =$ distance travelled
$t =$ time taken
$v =$ velocity.

Then $v = \frac{s}{t}$,

whence $s = vt$.

I.e. the distance passed over is equal to the number of units of distance multiplied by the number of units of time taken for the distance.

In using the above formula, care must be taken to employ correct units.

Example. *A train running uniformly passes over* 2 *km in* 4 *min* 10 *s. What is its speed?*

Using $v = \frac{s}{t}$ and substituting given values:

$$v = \frac{2000\text{ m}}{4\text{ min }10\text{ s}}$$
$$= \frac{2000\text{ m}}{250\text{ s}}$$
$$= 80\text{ m/s}.$$

90. Average velocity

When the velocity or speed is not uniform we can obtain the average velocity by dividing the total distance passed over by the total time taken.

If, therefore, v represents the average velocity, and s and t the total distance and time, then, as before, the formula $v = \frac{s}{t}$ still holds; but in the case of variable velocity v represents the average velocity for the specified distance.

This may be illustrated by a railway example.

Example. *A train which leaves Euston at* 10.50 *a.m. arrives at Manchester at* 3.10 *p.m. What is the average velocity for the journey* 335 *km?*

Time taken is 1 h 10 mins + 3 h 10 mins = 4 h 20 mins.

∴ Using

$$v = \frac{s}{t}, \text{ where } v = \text{average velocity},$$

$$v = \frac{335 \text{ km}}{4 \text{ h } 20 \text{ mins}} = \frac{335}{48}$$

$$= 77{\cdot}5 \text{ km/h nearly.}$$

91. Distance–time graphs

The motion of a body, whether uniform or variable, can be represented graphically. The following examples illustrate the method.

(1) Uniform motion

A motor-car passes over distances in times as shown in the following table. Exhibit them as a graph.

Time in s	1	2	3	4	5
Distances in m from starting point	2	4	6	8	10

Any of these distance divided by the corresponding interval of time gives $v = 2$ m/s uniformly. Fig. 114 shows the above quantities plotted,

time in seconds along OX.
distances in metres along OY.

Joining the plotted points, we see that they lie on a straight line.

This is the **distance–time** graph for the motion described in the table.

If any point P be taken and PQ be drawn perpendicular to OX, the ratio $\frac{PQ}{OQ}$ is constant.

This ratio is called the **gradient** of OP,

and $$\text{velocity} = \frac{\text{distance}}{\text{time}} = \frac{PQ}{OQ}.$$

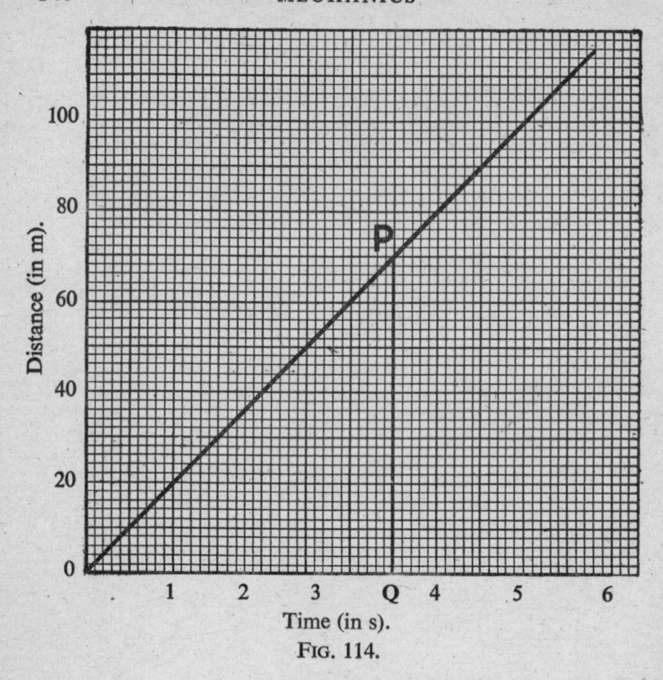

FIG. 114.

Thus the **gradient of the straight line *OP* measures the velocity.** It will be noted that when velocity is uniform the distance–time graph will always be a straight line.

(2) Velocity varying uniformly

The following table shows the distances passed over by a falling body.

Time (in s) . .	0	0·2	0·4	0·6	0·8	1·0	1·2	1·4	1·6
Distance (in m) .	0	0·2	0·79	1·76	3·14	4·90	7·05	9·60	12·5

Representing time along OX
and distance along OY,

the values of the table are plotted as shown in Fig. 115.

When the points are examined they are seen to lie on a smooth curve which is part of a parabola.

The velocity thus changes from point to point.

The curve becomes steeper as the time increases. If the increases in distance over corresponding times are cal-

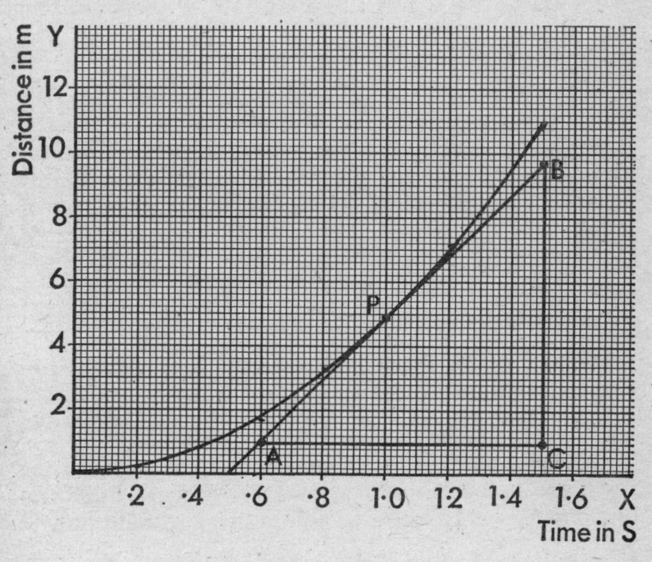

Fig. 115.

culated it is obvious that the velocity is increasing. Remembering the conclusions above, it is reasonable to conclude that, if we wish to know the velocity at any particular instant, we should draw the tangent to the curve at the corresponding point. The gradient of the tangent shows the way in which the distance is increasing for a tiny increase in time at that point. In other words, the gradient of the tangent is the velocity at the point.

Thus to find the velocity a second after the motion has begun—i.e. at the point P on the curve—a tangent to the curve should be drawn at P.

The gradient of this tangent, APB, is given by $\frac{BC}{AC}$ and this represents the velocity at P.

Substituting values we have

$$\frac{BC}{AC} = \frac{9{\cdot}8 - 1{\cdot}0}{1{\cdot}5 - 0{\cdot}6} = \frac{8{\cdot}8}{0{\cdot}9}.$$

Thus the velocity at the end of the first second is $9{\cdot}\dot{8}$ m/s.

When the velocity is varying uniformly the graph will be a regular curve and **the velocity at any point will be given by the gradient of the tangent to the curve at that point.**

92. Velocity–time graphs

Graphs which show how velocity is changing with respect to time in the motion of a body are called velocity–time graphs.

(1) The case of uniform velocity

Let us consider the velocity–time graph for a body moving with a uniform velocity of 6 m/s (Fig. 116).

This means that no matter what the interval of time, whether it be 1 s or 5 s, the velocity will be the same—i.e. 6 m/s.

∴ The distance of any point on the graph from OX will always be the same—i.e. the distance corresponding to 6 on the velocity axis.

The graph must therefore be one straight line AP, which is parallel to OX and 6 units distant from it. We may conclude:

If a body is moving with uniform velocity, its velocity–time graph will be a straight line parallel to OX.

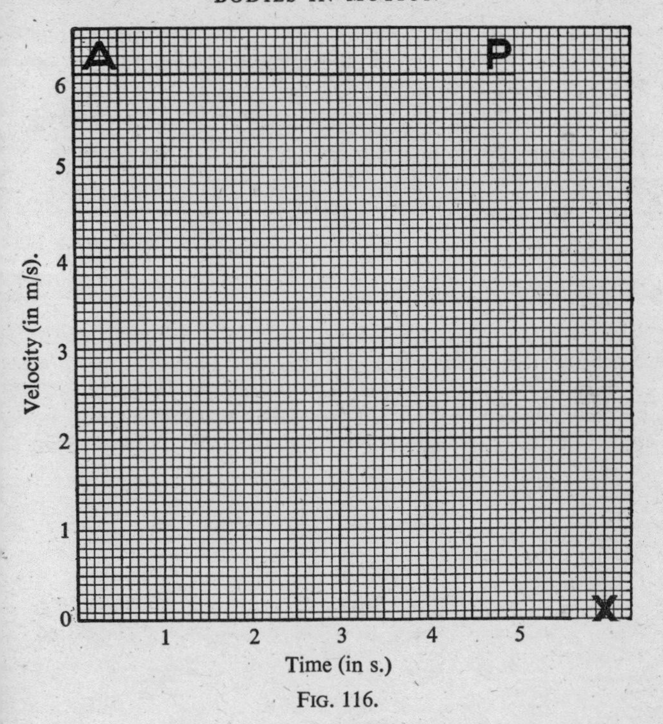

FIG. 116.

(2) Velocity increasing or decreasing uniformly.

(*a*) **Velocity increasing uniformly.** Let the velocity of a body increase uniformly by 2 m/s each second.

The following table shows the velocity after each second.

Time (in s) . .	0	1	2	3	4
Velocity (in m/s) .	0	2	4	6	8

Fig. 117 shows the velocity–time graph of the body.

The graph is constructed by obtaining points which show the velocity at the end of each second. These points, as might be expected, lie on a straight line.

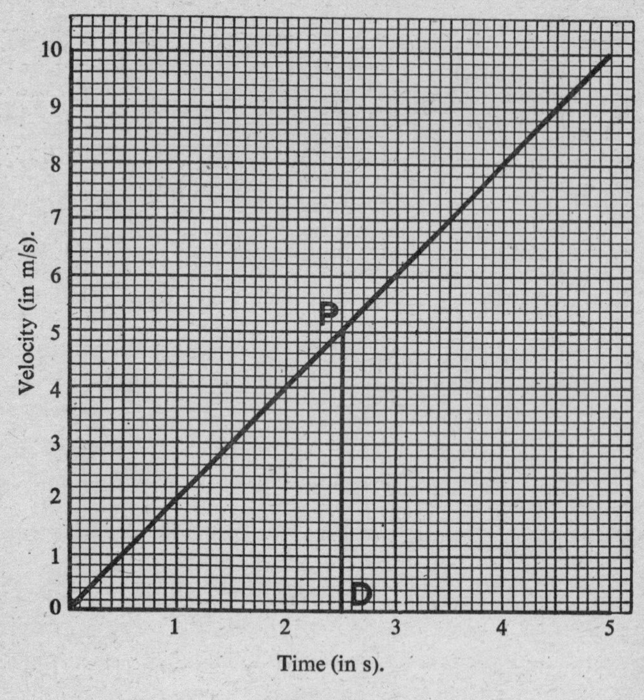

FIG. 117.

This straight line is the velocity–time graph.

If **any** point, P, be taken on the graph and a perpendicular PD be drawn to OX, then the ratio $\frac{PD}{OD}$ is constant wherever P is taken. The graph is one of constant gradient and therefore a straight line. This gradient represents the increase in velocity per second.

(*b*) **Velocity decreasing uniformly.** Fig. 118 is the velocity–time graph of a body having an initial velocity of 7 m/s, which decreases each second by 1 m/s.

The decrease being uniform, the points, which mark the velocity after each second, will be found to lie on

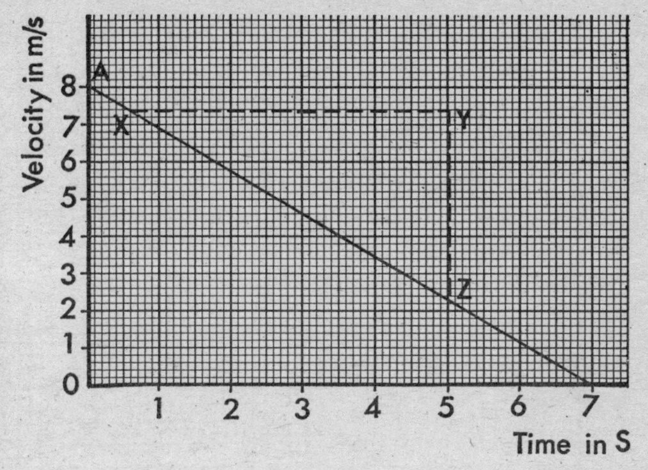

FIG. 118.

the straight line AC. The gradient of the line in this case is *negative.*

$$\text{Gradient} = \frac{\textit{increase} \text{ in velocity}}{\text{time taken}},$$

but in this case we have a *decrease.* A decrease is the same thing as a negative increase. In the 6 seconds shown, for example, the velocity has "increased" by "minus 7 m/s".

Thus the gradient is $\dfrac{-YZ}{XY}$.

(3) **Velocity–time graph of a train** (velocity not uniform).

Fig. 119 represents a rough approximation of the velocity–time graph of an electric train.

The train starts from rest at O, and its velocity increases rapidly and uniformly until in 2 mins it reaches 30 km/h at A.

It then travels **uniformly** at 30 km/h for 22 mins shown at B. Afterwards its velocity decreases rapidly until the trains comes to rest at C after $1\frac{1}{2}$ mins.

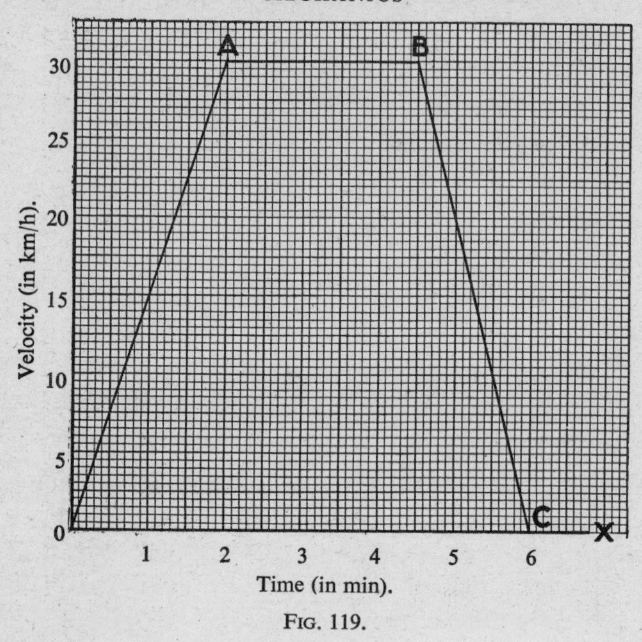

FIG. 119.

93. Area under a velocity–time graph

Example 1

Fig. 120, which is Fig. 116 reproduced on opposite page for convenience, is the velocity–time graph for a body moving for 5 s with a uniform velocity of 6 m/s. AP, parallel to OX, is the graph.

From any point P draw PQ perpendicular to OX. The area of the rectangle $OAPQ$ is called **the area under the graph.**

Area of $OAPQ = OA \times OQ$.

Now OA = number of m/s, i.e. velocity, travelled throughout the time

and OQ = number of seconds taken to move from A to P.

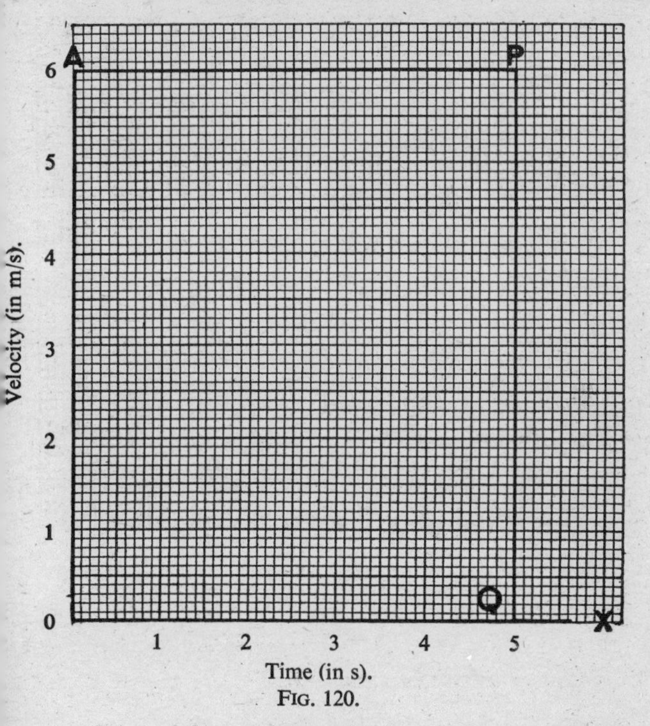

FIG. 120.

$\therefore$ $OA \times OQ$ = number of units of velocity $\times$ number of units of time,
= 6 m/s $\times$ 5 s
= 30 m.

But $OA \times OQ$ = the area of the rectangle $OAPQ$.

$\therefore$ **Number of units of distance passed over in the time by the body = number of units of area in *OAPQ*.**

Example 2

Assuming, without further experiment, the truth of the conclusion reached in the last paragraph, we will apply it to find the distance passed over by the train whose velocity–time graph was shown in Fig. 119.

There is, however, one important difference in this case.

In Fig. 120 the units of time were the same on the two axes, and this was an important point in the argument.

But in Fig. 121 the unit of time employed in the velocity scale marked on OY is the hour, while that employed in the time scale on OX is the minute.

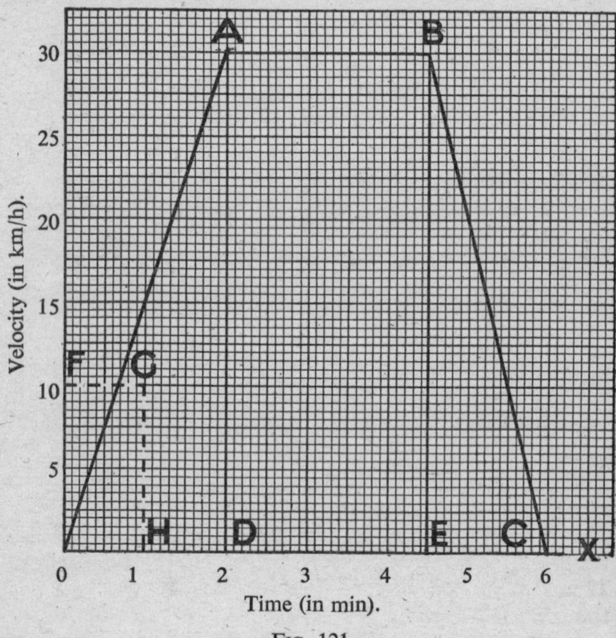

FIG. 121.

It is important, therefore, to begin by ascertaining what is represented by each unit of area in the figure.

We will take the rectangle *OFGH* as a unit and see what that represents. *OFGH* represents a velocity of 10 km/h, acting for one *minute*.

We must therefore express the minutes in hours.

$\therefore$ *OFGH* represents 10 km/h for $\frac{1}{60}$ h $= \frac{1}{6}$ km.

We now find the area of the whole expressed in terms of rectangle *OFGH* as a unit.

We may either find the area of the trapezium $OABC$ by using the ordinary formula, or, by drawing AD and BE, we can find separately the area of the triangles AOD, BEC and the rectangle $ABED$. Regarding rectangle $OFGH$ as a unit:

Area under:

$$AO = 3 \text{ units} = 3 \times \tfrac{1}{6} \text{ km} = \tfrac{1}{2} \text{ km.}$$
$$AB = 7\tfrac{1}{2} \text{ units} = 7\tfrac{1}{2} \times \tfrac{1}{6} \text{ km} = 1\tfrac{1}{4} \text{ km.}$$
$$BC = (\tfrac{1}{2} \times 4\tfrac{1}{2}) \text{ units} = 2\tfrac{1}{4} \times \tfrac{1}{6} \text{ km} = \tfrac{3}{8} \text{ km.}$$

$\therefore$ Total distance $= \frac{1}{2} + 1\frac{1}{4} + \frac{3}{8} = 2\frac{1}{8}$ km.

As was stated in § 92, the graph $OABC$ is a simplified approximation to the real graph. In reality the curve would be an irregular one. The methods for finding the areas in such cases can be studied in books on "Practical" mathematics, such as *National Certificate Mathematics.*

Exercise 13

1. If a train travels 400 m in 12 s, what is its speed in km/h?
2. If a bicycle is travelling at 24 km/h, how many metres will it go in 15 s?
3. A man travelling in a train notices that he passes a telegraph pole every 5 s; he knows that the poles are 125 m apart. What speed does he (correctly) deduce that the train is travelling at?
4. The speed of light may be taken as $2{\cdot}997\,925 \times 10^8$ m/s. What is this in km/h?
5. A train is moving at 14 m/s. Its speed is increased each second by 0·5 m/s. What will its speed in km/h be after a minute?
6. A train is moving at 110 km/h, and is brought smoothly to rest in 10 s. At what rate per second is its speed decreased?
7. A train is travelling at 12·2 km/h and its speed is increased in each second by 3 m/s. Another train is moving at 19·4 km/h and its speed is increased in each

second by 2 m/s. In how many seconds will the trains be travelling at the same speed and, in km/h, what will that be?

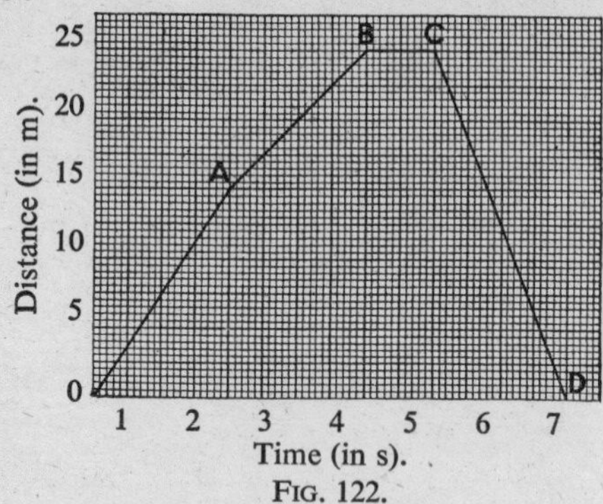

FIG. 122.

8. Fig. 122 is the displacement–time graph of a body in motion. Find from it the average speed from O to A, A to B, B to C and C to D.

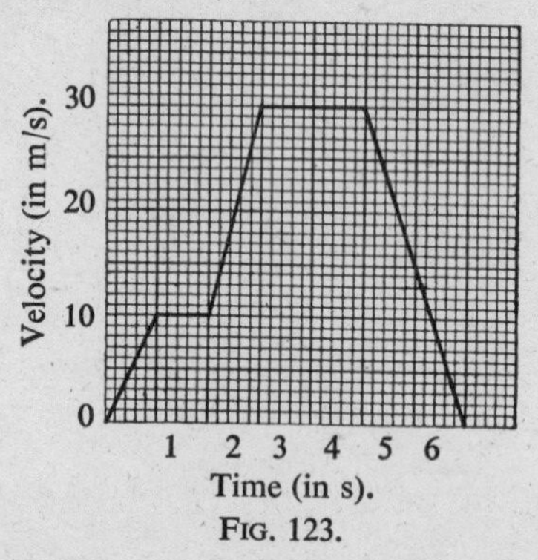

FIG. 123.

9. The distances travelled by a body at intervals of 5 min were as shown in the following table:

Time (in min) . .	5	10	15	20	25	30
Distance (in km) .	10	15	19	22	24	25

Plot the distance–time graph and find the average speed over each of the six intervals.

10. Fig. 123 shows the speed–time graph of a body moving with speeds as shown at the end of successive seconds. Find from the graph the total distance travelled by the body.

CHAPTER IX

ACCELERATION

94. Changes in velocity

In the previous chapter cases were considered in which the velocity of a body was increased or decreased.

If the velocity of a body is increased, the motion is said to be accelerated; if the velocity is decreased, the motion is said to be retarded.

It is very important that we should examine the **rate** at which the motion is being changed.

The rate at which the velocity is increased is called the acceleration of the body.

Thus, if the velocity of a body is being uniformly increased by 3 m/s in every second, the acceleration is said to be 3 m per second, per second. The use of "per second" twice is sometimes puzzling to beginners, but a little consideration will show the need for the use of them. The first "per second" refers to the rate of the velocity and was used in this way in the previous chapter. The second "per second" gives the unit of time in which the additional amount of velocity is added on.

An acceleration of 3 m per second per second is usually abbreviated into 3 m/s^2, where the index shows that the "per second" is repeated.

We shall make little use of the term "retardation". The term "acceleration" is used to mean that there may be an increase or a decrease. If there is an increase the sign of the quantity is positive, if there is a decrease the sign is negative.

Thus an acceleration of -5 m/s^2 means that the velocity is **decreasing** by 5 m/s in every second.

Other units may be used. Thus we may speak of an acceleration of 2 km per hour per hour or 2 km/h^2.

95. Uniform acceleration

When the velocity of a body is increased by the same amounts in equal intervals of time, the acceleration is uniform. The cases we shall consider will be generally those of uniform acceleration.

96. Formula for a body whose motion is uniformly accelerated

Let the velocity of a body be 10 m/s.
Let it have a uniform acceleration of 3 m/s^2.
Then velocity after 1 second

$$= 10 \text{ m/s} + (1 \times 3) \text{ m/s}$$
$$= 13 \text{ m/s.}$$

Velocity after 2 seconds
$= 10 \text{ m/s} + (2 \times 3) \text{ m/s} = 16 \text{ m/s.}$
Velocity after 3 seconds
$= 10 \text{ m/s} + (3 \times 3) \text{ m/s} = 19 \text{ m/s.}$
$\therefore$ Velocity after t seconds
$= 10 \text{ m/s} + (t \times 3) \text{ m/s} = (10 + 3t) \text{ m/s.}$
From inspection of these results a general formula may be readily constructed.
Let u be the initial velocity of a body in m/s.
Let a be the acceleration in m/s^2.
Then

Velocity after 1 second $= (u + a)$ m/s
Velocity after 2 seconds $= (u = 2a)$ m/s
Velocity after 3 seconds $= (u + 3a)$ m/s.

And generally

Velocity after t seconds $= (u + (t \times a))$ m/s.

Let v be the velocity after t seconds.

Then $\qquad v = (u + at)$ m/s.

The same formula will be true whatever the units employed, provided they are the same throughout. Therefore we may state the formula generally as:

$$\boldsymbol{v = u + at}$$

from which $$t = \frac{v - u}{a}.$$

Also $$a = \frac{v - u}{t}.$$

97. Average velocity and distance

Fig. 124 represents the general velocity time diagram of a body, with uniform acceleration.

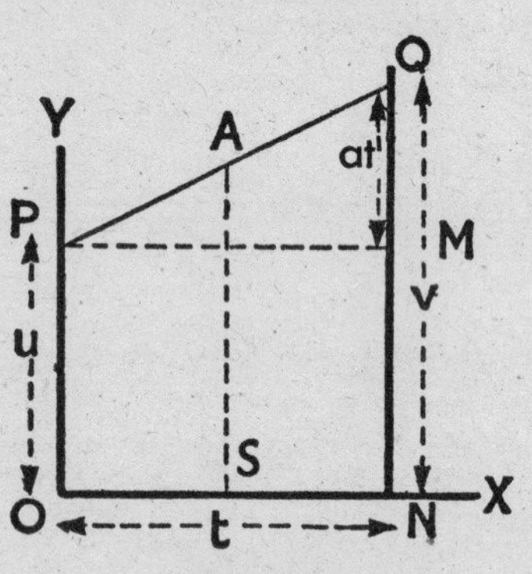

FIG. 124.

Let
$$u = \text{initial velocity}$$
$$v = \text{final velocity}$$
$$a = \text{acceleration}$$
$$t = \text{time.}$$

In the figure

OP represents u
QN represents v
ON represents t.

Draw PM parallel to OX.
Then QM represents $a \times t$ increase in velocity.

Also
$$QN = MN + QM$$
$$= OP + QM$$
or
$$v = u + at.$$

Let A be the mid point of PQ.
Draw AS perpendicular to ON.

Now, as shown in the previous chapter:

Distance passed over by the moving body is represented by the area under the graph.

$\therefore$ Distance passed over in the above case is represented by the area of the trapezium *OPQN*.

Using the formula for the area of a trapezium:

$$\text{Area of } \quad OPQN = \tfrac{1}{2}(OP + NQ) \times ON$$
$$= \tfrac{1}{2}(u + v) \times t.$$

$\therefore$ If s = distance passed over

$$S = \frac{u + v}{2} \times t.$$

Also, we know from Geometry that

$$AS = \tfrac{1}{2}(OP + QN)$$
$$= \tfrac{1}{2}(u + v).$$

$\therefore$ Average of initial and final velocities
= velocity at the middle of the interval.

$\therefore$ **S = (average of initial and final velocities) $\times$ t.**

Also **S = (velocity at the middle of the interval) $\times$ t.**

98. Distance passed over by a uniformly accelerated body

Let u, v, t, s represent the same quantities as in the previous paragraph.

As shown above,

$$\frac{u + v}{2} = \text{average velocity} \quad . \quad . \quad . \quad (1)$$

$$s = \frac{u + v}{2} \times t \quad . \quad . \quad . \quad . \quad . \quad . \quad (2)$$

and
$$v = u = at \quad . \quad . \quad . \quad . \quad . \quad . \quad . \quad (3)$$

In (2) for v substitute $u = at$ from (3).

Then
$$s = \frac{u + (u + at)}{2} \times t$$
$$= \frac{2u + at}{2} \times t$$
$\therefore$
$$s = ut + \tfrac{1}{2}at^2.$$

Again from (2)

$$s = \frac{u + v}{2} \times t.$$

$$\therefore \qquad u + v = \frac{2s}{t}.$$

Also $\qquad v - u = at.$

Multiplying $\qquad \boldsymbol{v^2 - u^2 = 2as,}$

whence $\qquad \boldsymbol{s = \frac{v^2 - u^2}{2a}.}$

This gives a formula for the distance in terms of the initial and final velocities.

Collecting the formulæ

$$(1) \quad \boldsymbol{v = u + at.}$$
$$(2) \quad \boldsymbol{s = \left(\frac{u + v}{2}\right) \times t.}$$
$$(3) \quad \boldsymbol{s = ut + \tfrac{1}{2}at^2.}$$
$$(4) \quad \boldsymbol{v^2 - u^2 = 2as.}$$

If a body starts from rest, then $u = 0$ and we get:

$$(5) \quad \boldsymbol{v = at.}$$
$$(6) \quad \boldsymbol{s = \tfrac{1}{2}at^2.}$$
$$(7) \quad \boldsymbol{v^2 = 2as.}$$

99. Retardation

When a body is uniformly retarded, a is replaced by $- a$, and the formulæ above which contain a become:

$$\boldsymbol{v = u - at.}$$
$$\boldsymbol{s = ut - \tfrac{1}{2}at^2.}$$
$$\boldsymbol{v^2 = u^2 - 2as.}$$

100. Worked examples

Example 1. *A train moving at* 36 *km/h is uniformly accelerated so that after* 8 *s its velocity is* 54 *m/s. Find the acceleration in m/s*^{2} *and the distance the train goes in the interval.*

Here
$$u = 36 \text{ km/h}$$
$$= 36 \times \frac{10}{36} \text{ m/s}$$
$$= 10 \text{ m/s.}$$
$$v = 54 \text{ km/h}$$
$$= 54 \times \frac{10}{36} \text{ m/s}$$
$$= 15 \text{ m/s.}$$

To find acceleration

Using $v = u + at$.

Then $15 = 10 + 8a$

$\therefore$ $a = \frac{5}{8} \text{ m/s}^2$.

To find distance

Using
$$s = \frac{u + v}{2} \times t,$$
$$s = \frac{10 + 15}{2} \times 8$$
$$= 100 \text{ m.}$$

Alternatively we could have used either $s = ut + \frac{1}{2}at^2$
or $v^2 - u^2 = 2as$.

The student is advised to use these to check that the answer is correct.

Example 2. *A body has an initial velocity of* 18 *km/h and is uniformly accelerated at a rate of* 9 *cm/s*^{2} *over a distance of* 200 *m. Find, in m/s, the speed it acquires.*

$$u = 18 \text{ km/h} = 18 \times \frac{10}{36}$$
$$= 5 \text{ m/s.}$$
$$a = 9 \times 10^{-2} \text{ m/s}^2.$$

To find final velocity

Using $v^2 - u^2 = 2as$,
$$v^2 = u^2 + 2as$$
$$= 25 + 2 \times 9 \times 10^{-2} \times 200$$
$$= 25 + 36$$
$$= 61$$
$$\therefore \quad v = \sqrt{61}$$
$$= 7{\cdot}8 \text{ m/s.}$$

Example 3. *The brakes on a train reduce its speed from 72 km/h to 54 km/h while it moves 100 m. Assuming that they cause a uniform retardation, find how much farther the train will run before coming to rest, and how long it will take.*

$$u = 72 \times \frac{10}{36} = 20 \text{ m/s.}$$

$$v = 54 \times \frac{10}{36} = 15 \text{ m/s.}$$

To find the "acceleration"

Using $$v^2 - u^2 = 2as,$$

$$15^2 - 20^2 = 2 \times 100a$$

$$\therefore \quad (15 + 20) \times (15 - 20) = 200a$$

$$\therefore \quad \frac{35 \times 5}{200} = a.$$

$$\therefore \quad a = -\frac{7}{8} \text{ m/s}^2.$$

N.B. The minus sign shows that really it is not an acceleration but a retardation.

To find the extra distance run

Using $$v^2 - u^2 = 2as,$$

$$v = 0; u = 15 \text{ m/s}$$

$$\therefore \quad 0 - 15^2 = 2 \times \left(-\frac{7}{8}\right) \times s$$

$$\therefore \quad \frac{15^2 \times 8}{2 \times 47} = s$$

$$\frac{900}{7} = s$$

$$\therefore \quad s = 129 \text{ m approx.}$$

To find the time taken

Using $$v = u + at,$$

$$0 = 15 - \frac{7}{8}t$$

$$\therefore \quad t = \frac{120}{7}$$

$$= 17 \text{ s approx.}$$

Exercise 14

1. A body starts from rest and moves with an acceleration of 5 m/s^2. What will be its speed after 8 s, and how far will it have gone?

2. How long would a body moving under the conditions of the previous question take to travel 1 km?

3. A car is moving at 40 km/h. If brakes are put on and the car is retarded at the rate of 4 m/s^2, after how long will it come to rest, and how far will it travel in doing so?

4. A train starting from a station reaches a velocity of 36 km/h in 40 s. Find the acceleration, and the distance travelled by the train in the 40 s.

5. A train travelling at 48 km/h is brought to rest by its brakes in 20 s. What is its retardation, and how far does the train travel in that time?

6. The speed of a car is increased from 40 km/h to 60 km/h in travelling 200 m. What was the acceleration of the car, and how long did it take to travel the distance?

7. The velocity of an aeroplane was increased from rest by 3 m/s^2 for 25 s. It was then retarded at 2·5 m/s^2 until it came to rest. How far did it travel in all?

8. A body starting from rest has a constant acceleration of 4 m/s^2. How far will it travel in the 10th second?

9. A ship is moving at 15 km/h. If it has to pull up in 4 km, what must be its retardation?

10. A car travelling at 40 km/h has its velocity reduced to 10 km/h in 2·5 s. What distance was covered in that time?

11. A train moves from rest with uniform acceleration of 1·5 m/s^2. After what time and distance will its speed be 100 km/h?

12. A slip carriage is detached from a train moving at 60 km/h. If it comes to rest in 625 m, how long is this after the carriage left the train?

13. A body has an initial velocity of 150 m/s and an acceleration of 120 m/s^2. How long will it take to travel 7·5 km?

14. The brakes on a vehicle reduce its speed to 20 km/h from 60 km/h, while it moves 125 m. Assuming that the

brakes exert a constant retarding force, how much farther will the vehicle continue before coming to rest, and how long will it take?

101. Acceleration due to gravity

The most common example which we experience of accelerated velocity is that of gravity. It had been observed in the earliest days of the human race that if a body fell from a height to the ground, the farther it fell, the greater was its velocity on reaching the ground. Speculation about it, however, was mainly concerned with the question: Why does the body fall? To this problem the answer has not yet been found. The great mathematician, Galileo, seems to have been the first to consider the question: *How* does it fall? That is, what is the nature of the motion itself? Is it uniformly accelerated? What is the acceleration? These are the kind of questions that Galileo set himself to answer. He left for future generations the problem of the cause of the motion.

Before the time of Galileo it had been generally accepted that a heavy body fell to the ground with greater velocity than a light body. This was superficially deduced from the fact that such an object as a feather would flutter slowly to the ground, whereas a piece of stone falls rapidly.

Tradition has it that Galileo performed the experiment of dropping bodies of different weights from the top of the leaning tower of Pisa, and showed that they reached the ground in approximately the same time. But his experiments and deductions did not convince the doubters and he was expelled from the University of Pisa.

In the next century Sir Isaac Newton showed that the resistance of the air accounted for the differences in the times that certain bodies took in falling by performing his famous experiment of dropping a feather and a guinea in a cylindrical glass vessel from which nearly all the air had been exhausted. He showed that both feather and coin fell in the same time.

102. The acceleration of a falling body

Since the time of Galileo various forms of apparatus have been devised to show that bodies fall with uniform acceleration and to discover what is the amount of the acceleration. The student who has access to a physical laboratory will be able to study these experiments, but in this book we must be content with stating what are the facts which they demonstrate. By means of them we learn that—

The acceleration of a falling body is uniform and is approximately 9·8 m/s².

The letter g is always used to denote this acceleration.

The value of g is given above as approximately 9·8 m/s², but it varies slightly at different places on the earth's surface, for reasons which will not be considered now. It has its greatest value at the poles, where it is approximately equal to 9·83, and its least value at the equator, where it is 9·78, the figure again approximate.

At Greenwich the value of g in vacuo, i.e. the air resistance eliminated, is 9·806 65 m/s², correct to five places of decimals; generally in this book we shall follow the custom of assuming the value of 9·8 m/s².

103. Formulæ for motion under gravity

The formulæ proved in § 98 were general and hold for any uniform acceleration. They are therefore true for falling bodies, but it is usual, when employing them, to replace a by g. With this substitution the formulæ are as follows:

$$(1)\quad v = u + gt.$$
$$(2)\quad s = \left(\frac{u + v}{2}\right) \times t.$$
$$(3)\quad s = ut + \tfrac{1}{2}gt^2.$$
$$(4)\quad v^2 = u^2 - 2gh.$$

If a body is **dropped** from a height, it starts with zero velocity, and consequently the formulæ are modified as follows:

$$(5)\quad v = gt.$$
$$(6)\quad s = \tfrac{1}{2}gt^2.$$
$$(7)\quad v^2 = 2gh.$$

If g be taken as equal to 9·8, then (5) and (6) become:

$$v = 9{\cdot}8t.$$
$$s = 4{\cdot}9t^2.$$

These are useful forms to remember.

104. Motion of a body projected vertically upwards

When a body is projected upwards, the motion is retarded, i.e. the acceleration is negative. It will rise until it reaches its maximum height, when for an instant the velocity is zero. If v is the velocity of projection, then time to reach this is given by

$$t = \frac{v}{g}.$$

If h be the greatest height, then, using formula 7 above,

$$h = \frac{v^2}{2g}.$$

At the maximum height the motion of the body is reversed and the body begins to fall under the action of gravity, i.e. with acceleration g.

Consequently, if there were no air resistance, it would reach the ground again in the time it took to rise and with the velocity with which it was projected, but in the opposite direction.

The circumstances of the downward path are those of the upward reversed.

The total time from ground to ground will be

$$2t \text{ or } \frac{2v}{g}.$$

The total distance will be

$$2h \text{ or } \frac{v^2}{g}.$$

105. Worked example

A bullet is shot vertically upwards with a velocity of 49 m/s from a stationary balloon, 98 m above the ground. Find:

(1) *The greatest height and the time taken to reach it.*

(2) *Time to reach the ground, assuming no interception by the balloon.*

(3) *The velocity on reaching the ground.*

(Air resistance is to be ignored throughout.)

(1) **Time and greatest height.**

Using the formula

$$t = \frac{v}{g}$$

$$t = \frac{49}{9{\cdot}8} = 5 \text{ s.}$$

Height.

Using $s = 4{\cdot}9t^2$,

we have $s = 4{\cdot}9 \times 5^2 = 122{\cdot}5$ m (above the balloon).

(2) From the greatest height the bullet has to fall, viz. (dist. above the balloon) + (dist. of balloon above ground)

$$= 122{\cdot}5 \text{ m} + 98 \text{ m} = 220{\cdot}5 \text{ m.}$$

Using $s = \frac{1}{2}gt^2$, $220{\cdot}5 = 4{\cdot}9t^2$

$$t^2 = 45$$

$$t = \sqrt{45} = \mathbf{6{\cdot}7\ s} \text{ (approx.).}$$

This is the time from the highest point, but it took 5 s to rise to that.

$\therefore$ Total time $= 6{\cdot}7 + 5 = 11{\cdot}7$ s.

(3) **Velocity on reaching the ground.**

Using $v^2 = 2gh$,

we have $v^2 = 2 \times 9{\cdot}8 \times 220{\cdot}5$

$= 4322$

$\therefore$ $v = \sqrt{4322}$

$= 65{\cdot}7$ m/s.

An alternative method of approach would have been to use $v = gt$.

Exercise 15

Note. The effect of air resistance is to be neglected.

1. In the following table fill in the velocities and distance passed over by a body falling from rest ($g = 9{\cdot}8$ m/s^2).

Time (s) . . .	0	0·5	1	1·5	2	2·5
Velocity (m/s) . .	0					
Distance (m) . .	0					

From the results draw:

(1) A time–distance graph.
(2) A velocity–time graph.

Use the method of area under the graph to find the distance passed over in 2 s.

2. A stone is dropped from the top of a building and reaches the ground after 2·25 s. How high is the building ($g = 10$ m/s^2)?

3. How long does it take a falling body to acquire a velocity of 100 m/s, and through what distance does it fall in that time ($g = 9{\cdot}8$ m/s^2)?

4. A body is projected vertically upward with a velocity of 49 m/s. Find the greatest height reached and the time taken ($g = 9{\cdot}8$ m/s^2).

5. A stone is thrown vertically upwards and returns to the ground after 5 s. Find the initial velocity and the greatest height reached ($g = 9{\cdot}81$ m/s^2).

6. What is the velocity of a body after it has fallen 10 m, and how long does it take ($g = 9{\cdot}8$ m/s^2)?

7. A stone is dropped from a stationary helicopter. How far does it fall in the 4th second ($g = 9{\cdot}8$ m/s^2)?

8. A stone is projected vertically upwards with a velocity of 28 m/s. After what times is it 32 m high ($g = 10$ m/s^2)?

9. A balloon is rising vertically with a speed of

40 m/s. A stone is dropped from the balloon and reaches the ground after 3 s. How high is the balloon when the stone lands ($g = 10$ m/s^2)?

10. A body is projected upwards with a speed of 25 m/s. What is its velocity when it has reached a height of 20 m? How long will this be after the moment of projection ($g = 10$ m/s^2)?

11. A balloon ascends from rest with uniform acceleration, and a stone is dropped from it after 6 s. The stone reaches the ground 4 s after being released. Find the acceleration of the balloon and the height from which the stone fell ($g = 9{\cdot}8$ m/s^2).

12. Find the greatest height attained by a projectile which is thrown vertically upwards with a velocity of 25 m/s, and also the time which elapses before it returns to its point of departure ($g = 9{\cdot}8$ m/s^2).

CHAPTER X

NEWTON'S LAWS OF MOTION

106. Motion without forces

We now turn to the way in which forces produce movement. In Chapter I it was stated:

A single force will cause a body to accelerate.

For some readers this must have seemed strange, for many people believe that a single force is necessary to keep a body moving, let alone to accelerate it. The reason for this is that usually there is some friction. Bodies slow down on the earth because of the retarding forces provided by friction. A skier has a frictional force on his skis; a car has frictional forces between all its moving parts, and is additionally slowed down by wind resistance.

Such a situation is peculiar to places where the conditions are similar to those on earth, however. Once a rocket has escaped the pull of gravity from the earth there is no need to continue to use the rocket motors to keep it moving forwards. In the absence of the pull of gravity, air resistance and other frictional forces, it would go on and on and on—for ever. This is summed up in the following law.

107. Newton's First Law

A body remains in a state of rest or of uniform motion in a straight line provided that there is no resultant force acting on it.

There is an important corollary to this law, namely:

If a body is not in a state of rest or of uniform motion in a straight line, there must be a resultant force acting on it.

We have already made use of this corollary, in Chapter I, when we deduced that, since bodies accelerate towards the earth when they are released, there must be a force of gravity.

108. Mass and inertia

It was in Chapter I also that we had an initial look at mass. It was stated that the mass of a body is the property which measures its inertia. *Saying that a body has inertia is another way of saying that it is difficult to speed the body up or to slow it down.*

Bodies with large mass are more difficult to accelerate than bodies with small mass. It is equally true to say that bodies with large mass are harder to slow down. A lump of lead has a much greater mass than a lump of polystyrene of the same size. It would be much more difficult to speed the lead up, but once it had increased its speed it would be much more difficult to slow it down. *This effect is the same everywhere. It does not depend on gravity at all.* A lump of lead thrown at you in a spaceship would hurt just as much as a similar lump thrown at you on the earth. In other words your body would have to exert just as much force to slow it down in both cases.

When we talk about the mass of a body we are referring to its inertia.

109. The standard mass

A lump of platinum kept under very carefully controlled conditions near Paris is the standard mass with which all other masses are compared. The mass of this lump of platinum is called **1 kilogramme (kg).**

110. Comparing inertias

The inertia of a body is the difficulty we have in making it accelerate.

Suppose two objects are subjected to the same force, and that their accelerations are measured.

Let the first body accelerate at 6 m/s^2 and the second body accelerate at 3 m/s^2.

Quite clearly the second body has only been able to accelerate half as much as the first body. We say that it has twice as much inertia.

If the first body were a standard kilogramme we would say that the mass of the second body must be two kilogrammes—i.e. if a force makes 1 kg accelerate at 6 m/s^2 and it makes another body accelerate at 3 m/s^2, the mass of the other body must be 2 kg.

To compare the inertias of two bodies we must subject them both to the same force and then measure the accelerations produced.

If one accelerates twice as much as the other we say it has only half as much inertia. If it accelerates at a third of the rate of the other we say it has three times as much inertia, and so on.

In general, if the same force makes one body accelerate at a_1 m/s^2 and the second body to accelerate at a_2 m/s^2, the inertia of the second body is $\frac{a_1}{a_2}$ × inertia of the first body.

Since the mass of a body is a measure of its inertia, if the mass of the first body is m_1 kg the mass of the second body m_2 kg is

$$m_2 = \frac{a_1}{a_2} \times m_1 \; kg.$$

This method of comparing masses does not depend at all on the place where the experiment is carried out. It would work equally well on the moon, in a large space-ship, or on earth.

It will be readily appreciated by the reader that this is not a practical method for comparing masses. It is possible to devise experiments in which the effect of friction is almost entirely removed, but as a method for determining the mass of, say, a packet of potatoes it leaves much to be desired! The practical aspect of comparing masses is dealt with in Chapter XII.

111. Newton's Second Law of Motion

Newton stated this as follows:

The force on a body is proportional to the rate of change of motion of the body, and is in the direction of the change of motion.

By "motion" Newton meant the product of **mass** and **velocity,** which we usually call **momentum.** We shall consider momentum further in Chapter XI, but at present suffice it to say that a massive body moving slowly and a body of small mass moving very rapidly may be equally hard to stop.

Suppose a body of mass m starts at velocity v_1 and finishes with velocity v_2 after a time t.

According to Newton's Law

Force is proportional to rate of change of momentum.

$\therefore$ Force is proportional to $\dfrac{mv_2 - mv_1}{t}$

or $\dfrac{m(v_2 - v_1)}{t}$,

but $\dfrac{v_2 - v_1}{t} =$ acceleration, a.

$\therefore$ Force is proportional to $\boldsymbol{m} \times \boldsymbol{a}$.

We thus see that **if two bodies each are subjected to the same force, they will both have the same product of $m \times a$.**

Suppose two bodies of mass m_1 and m_2 are given accelerations a_1 and a_2 by the same force.

Then $$m_2 \times a_2 = m_1 \times a_1$$

or $$m_2 = m_1 \times \frac{a_1}{a_2}.$$

This result is, as we would expect, in accordance with our definition of what we mean by "mass".

112. A unit of force—the Newton

Since force, mass and acceleration are interlinked, we make use of the relationship in defining the unit of force.

A force of 1 Newton will cause a mass of 1 kilogramme to accelerate at 1 m/s^2.

Newton's Law states that force is proportional to $m \times a$.

$\therefore$ if

a mass of 1 kg is accelerated at 1 m/s^2 by 1 N
a mass of m kg is accelerated at 1 m/s^2 by m N
a mass of m kg is accelerated at a m/s^2 by $m \times a$ N.

Thus

$$\underset{\text{(in Newtons)}}{\text{Force}} = \underset{\text{(in kg)}}{\text{Mass}} \times \underset{\text{(in m/s}^2\text{)}}{\text{Acceleration}}$$

$$F = M \times a.$$

113. The Newton and the kgf

Until this point we have been using the kgf as our unit of force.

1 *kgf is the force of attraction felt by a mass of* 1 *kg when it is on the earth's surface.*

The effect of this force is to cause the object, to which it is applied, to accelerate towards the earth at g m/s^2, where g is the acceleration due to gravity at that place.

Now

$$\underset{\text{(in Newtons)}}{\text{Force on body}} = \underset{\text{(in kg)}}{\text{Mass}} \times \underset{\text{(in m/s}^2\text{)}}{\text{acceleration}}$$

$\therefore$ Force of gravity on 1 kg $= 1 \times g$ Newtons.

$\therefore$ **1 kgf $= g$ Newtons.**

The acceleration due to gravity varies from place to place, but is approximately 9·8 m/s^2 (*see* § 102). It is because of this variation in g that the kgf is an unsatisfactory unit of force, unless everyone agrees to measure it at places with the same value of g. In an earlier attempt to resolve the problem it was proposed that the kgf should be the force of attraction felt by a mass of 1 kg in a place where the acceleration due to gravity was 9·806 65 m/s^2. Most engineers and other scientists are now using the Newton, which is the internationally recommended unit of force.

Rule: To find the weight of a body at a given place if the mass of the body is known:

$$W = Mg \textbf{ Newtons (N)},$$

where W = Force of attraction from gravity, i.e. weight.
M = Mass in kg.
g = acceleration due to gravity at that place, in m/s^2.

114. Worked examples

Example 1. Take g as 9·8 m/s^2. *Find the weight of a mass of* 3 *tonnes on the earth.*

$$\begin{aligned} 3 \text{ tonnes} &= 3000 \text{ kg.} \\ \text{Weight} &= Mg \text{ Newtons.} \\ \therefore \quad \text{Weight} &= 3000 \times 9{\cdot}8 \\ &= 29\,400 \text{ N} \\ \text{or} \quad &= 30\,000 \text{ N approx.} \end{aligned}$$

Alternatively

$$\begin{aligned} \text{Weight} &= 3000 \text{ kgf.} \\ 1 \text{ kgf} &= 9{\cdot}8 \text{ N.} \\ \therefore \quad \text{Weight} &= 29\,400 \text{ N.} \end{aligned}$$

Example 2. *A force of* 50 *N acts on a body of mass* 500 *g. Find the acceleration.*

$$\begin{aligned} 500 \text{ g} &= 0{\cdot}5 \text{ kg} \\ F &= M \times a. \\ \therefore \quad a &= \frac{F}{M} \\ &= \frac{50}{0{\cdot}5} \\ &= 100 \text{ m/s}^2. \end{aligned}$$

Example 3. *A body weighs* 490 *Newtons. What force is needed to make it accelerate at* 5 *m/s*²*?*

$$\text{Weight of body} = M \times g.$$

$$\therefore \quad \text{Mass of body} = \frac{\text{Weight}}{g}$$

$$= \frac{490}{9{\cdot}8}$$

$$= 50 \text{ kg.}$$

$$\text{Force needed} = M \times a$$

$$= 50 \times 5$$

$$= 250 \text{ N.}$$

Example 4. *A body of mass* 200 *g is acted upon by a force of* 4 *N for a period of* 3 *s. Find the final velocity of the body if it starts from rest.*

$$200 \text{ g} = 0{\cdot}2 \text{ kg.}$$

Since $$F = Ma$$

$$a = \frac{F}{M}$$

$$= \frac{4}{0{\cdot}2}$$

$$= 20 \text{ m/s}^2.$$

Using $$v = at,$$

$$\text{after 3 seconds } v = 20 \times 3$$

$$= 60 \text{ m/s.}$$

Example 5. *A force of* 5 *kgf is applied to a mass of* 7 *kg. If it starts from rest how far will it have travelled in* 2 *seconds?*

$$1 \text{ kgf} = 9{\cdot}8 \text{ N.}$$

$$\therefore \quad 5 \text{ kgf} = 9{\cdot}8 \times 5$$

$$= 49 \text{ N.}$$

$$a = \frac{F}{M}$$

$$= \frac{49}{7}$$

$$= 7 \text{ m/s}^2.$$

Using $$s = \tfrac{1}{2} at^2,$$

$$\text{distance moved} = \tfrac{1}{2} \times 7 \times 4$$

$$= 14 \text{ m.}$$

Example 6. *A train of mass* 400 *tonnes has brakes that can bring it to rest from* 72 *km/h in* 100 *m. What is the minimum force that must be applied by the brakes?*

$$72 \text{ km/h} = 20 \text{ m/s}.$$

Using $$v^2 - u^2 = 2as,$$

$$0 - 20^2 = 2 \times 100 \times a$$

$$\therefore \quad a = -\frac{400}{200}$$

$$= -2 \text{ m/s}^2.$$

The minus sign indicates that it is a retardation.

$$400 \text{ t} = 400\,000 \text{ kg}$$

$$= 4 \times 10^5 \text{ kg}.$$

$$\text{Retarding force needed} = M \times a$$

$$= 4 \times 10^5 \times 2$$

$$= 8 \times 10^5 \text{ N}.$$

Example 7. *A truck of mass* 2 *tonnes is to be hauled up an incline of* 1 *in* 14. *What constant force, applied parallel to the plane, is required to move it at a constant acceleration of* 8 *cm/s*^{2}*?*

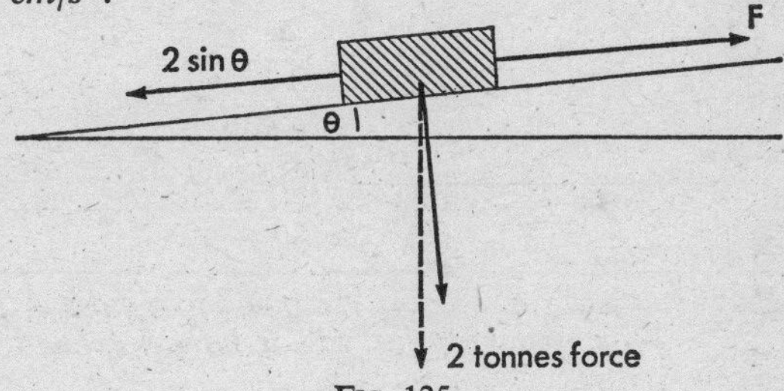

FIG. 125.

Let the force in Newtons acting parallel to and up the plane be F.

$$\text{Mass of the truck} = 2 \times 10^3 \text{ kg}$$

$$\therefore \quad \text{Weight of truck} = 2 \times 10^3 \times 9{\cdot}8$$

$$= 19{\cdot}6 \times 10^3 \text{ N}.$$

Resolved part of weight acting down the plane

$$= W \sin\theta \quad (\S\ 60)$$

$$= \frac{19{\cdot}6 \times 10^3}{14}$$

$$= 1{\cdot}4 \times 10^3 \text{ N}.$$

The forces acting on the truck parallel to the plane are:

$$F \quad \text{(up)}$$
$$1{\cdot}4 \times 10^3 \text{ Newtons (down)}$$
$$\therefore \quad \text{Resultant force} = F - 1400 \text{ N.}$$

The acceleration is 8 cm/s^2

$$= 8 \times 10^{-2} \text{ m/s}^2,$$

and

$$\begin{aligned} F &= Ma \\ &= 2 \times 10^3 \times 8 \times 10^{-2} \\ &= 160 \text{ N}, \\ \therefore \quad F - 1400 &= 160 \\ F &= 1560 \text{ N.} \end{aligned}$$

115. The Second Law and changing mass

Newton's Second Law does not only apply to situations in which a single body is accelerated. In jet propulsion, for example, the ejected gas has its momentum increased, but it is not the same gas the whole time. When a jet of water is stopped by a wall many different particles of water are retarded by the wall. The jet of water is continually losing momentum, and this implies that there must be a force to cause this.

Worked example

Water issues from the horizontal nozzle of a hose-pipe at a rate of 100 *g/s. The nozzle velocity of the water is* 150 *cm/s. If the water is stopped by a vertical wall, calculate the force that the wall must be exerting on the water to stop it.*

$$\begin{aligned} 150 \text{ cm/s} &= 1{\cdot}5 \text{ m/s} \\ 100 \text{ g} &= 0{\cdot}1 \text{ kg} \\ \text{"momentum" of } 100 \text{ g} &= m \times v \\ &= 0{\cdot}1 \times 1{\cdot}5 \\ &= 0{\cdot}15 \text{ kg m/s.} \\ \text{In 1 second momentum lost by water} &= 0{\cdot}15 \text{ kg m/s.} \\ \therefore \quad \text{Rate of change of momentum} &= 0{\cdot}15 \text{ kg m/s}^2. \\ \text{But "Rate of change of momentum"} &= \text{Force} \\ \therefore \quad \text{Force} &= 0{\cdot}15 \text{ N.} \end{aligned}$$

Exercise 16

(*Take g as* 9·8 m/s^2.)

1. A mass of 10 kg is acted on by a force of 4 N. What is the acceleration that is produced?

2. What force acting on a mass of 16 kg will produce an acceleration of 6 m/s^2?

3. A force of 10 N acting on a body produces an acceleration of 5 m/s^2. What is the mass of the body?

4. The speed of a car of mass 1 t increases from rest to 36 km/h in 8 s. If the force acting on the car was constant, what was it, in k N?

5. An engine exerts a pull of 110 kN on a train whose mass, including the engine, is 200 t. The resistance to motion totals 40 kN. What, in m/s^2, is the acceleration of the train?

6. If a car of mass 1·5 t, travelling at 45 km/h, is brought to rest over 15 m, what is the average force acting?

7. 10 kg, acted on by a constant force for 1 minute, increases its speed by 300 m/s. What is the magnitude of the force in the direction of motion?

8. A train of mass 240 t, running at 45 km/h, has its velocity reduced to 18 km/h in 12 s. What was the retarding force, in kN?

9. A shell, of mass 10 kg, falls for 3 s and then penetrates 0·4 m into a sand dune. What is the average resistance of the sand?

10. A car travelling at 40 km/h is brought to rest over 40 m. The car has mass 1500 kg. What was the average force acting, in N?

11. In launching a ship of 5000 t, it is necessary to stop her at a point 200 m from the end of the slipway. If the ship leaves the slip with a velocity of 4 m/s towards the point, what is the average force parallel to the slip which will stop the ship at the right point?

12. A trolley of mass 200 kg is pulled up a smooth incline of 1 in 10 with an acceleration of 2 m/s^2. What pulling force acts parallel to the incline?

116. Newton's Third Law of Motion

This is often stated as follows:

To every action there is an equal and opposite reaction.
An alternative way of expressing it is to say:

For every force acting on a body there is an equal and opposite force acting on another body.

Anyone who has kicked a medicine ball thinking that it was an ordinary football will be well aware of the fact that there was a force of retardation acting on his foot, but he may not have been aware that it was equal and opposite to the force on the ball.

117. Newton's Third Law and similar bodies

If two people sitting in rowing boats each hold one end of a rope, the boats are drawn together if one of them pulls the rope. So long as the tension in the rope is the same each time, the result of pulling is the same whichever person does the pulling, and is the same as if both pull. This situation is represented in Fig. 126. If T

FIG. 126.

Newtons is the tension in the rope each time, and the effective masses of boat + passenger are m_1 and m_2, the accelerations will be T/m_1 and T/m_2.

118. Newton's Third Law and dissimilar bodies

If a brick falls towards the earth we deduce that it is feeling a force. We call it the force of gravity. It is also true to say that the earth is feeling a force pulling it towards the brick. The effect is very small, of course, but we believe it to exist. It is this principle which is used in jet propulsion. The gases speed out backwards with increased momentum. This increased momentum is the result of a force applied by the engines. There is an equal and opposite force given to the engines, and hence to the aeroplane of which the engines are part, and it is this force which accelerates the plane. In exactly the same way the jet of water in § 116 produced a force on the wall.

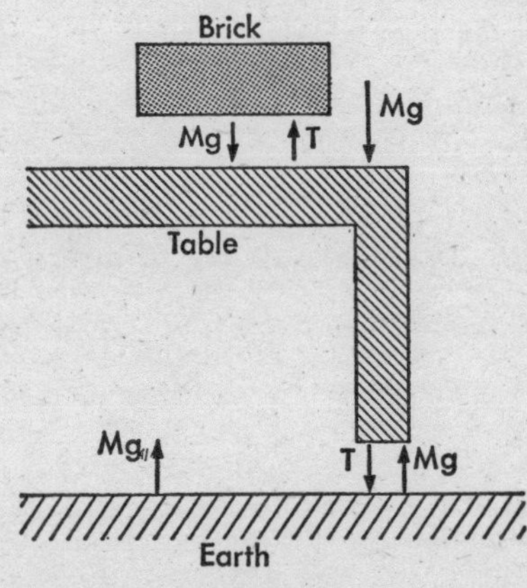

FIG. 127.

119. Newton's Third Law and statics

If a brick is placed on a table it is being pulled downwards onto the table by the force of gravity. Again there is an equal and opposite force on the earth, pulling the earth to the brick. In this case, however, the brick and the earth cannot approach each other, because the table is in the way. The effect of the forces, therefore, is to put the table in a state of compression. This is represented in Fig. 127.

The forces on the brick are:

Mg down.
T up, where T is the force
provided by the table trying to relieve its compression.

The forces on the table are:

Mg down, from the brick.
Mg up, from the earth.

The forces on the earth are:

Mg up, from the brick.
T down, from the table.

The situation will, in practice, be further complicated, because the table itself will be attracted by the earth, i.e. the table will have some weight.

120. Newton's Third Law and walking

When we start to walk or run we accelerate. There must, therefore, be a force to provide this. Where does the force come from? Where is the equal and opposite force on another body?

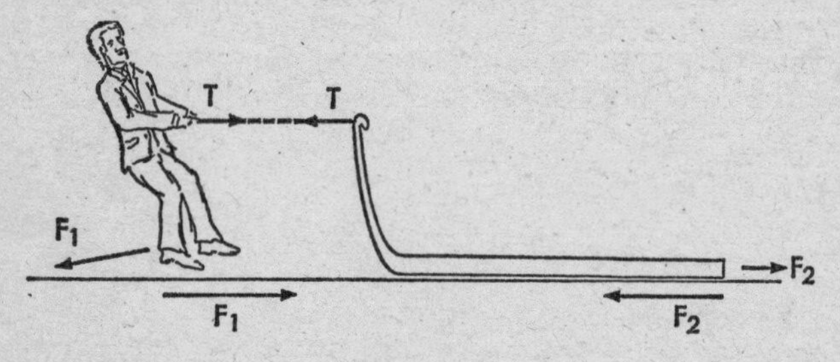

FIG. 128.

Quite clearly the agent enabling us to walk at all is the friction on the soles of our feet. Anyone who has tried to walk on a very slippy surface will testify to this fact. The force on the walker causes an equal and opposite force on the earth. This fact can easily be demonstrated if a person tries to run over a mat which has been placed on a polished floor. The mat is given a rapid acceleration backwards.

Whenever we show the forces acting on a body we must be very careful to show which body the forces are acting upon. Fig. 128 represents a man pulling a sledge.

The horizontal forces on the man are:

F_1 forwards, from the effect of the friction on his feet.
T backwards, which is the tension in the rope.

The horizontal forces on the sledge are:

T forwards.
F_2 backwards, from the frictional effect of the ground.

The horizontal forces on the ground are:

F_1 backwards.
F_2 forwards.

These latter forces can be demonstrated by trying to pull a heavy object over a carpet. The carpet under the man's feet is pushed back towards the object, while the object tries to drag the carpet with it.

If $F_1 > T$ the man accelerates forwards. Since the sledge moves with him, this also implies that $T > F_2$. Thus the condition for acceleration is that $F_1 > F_2$.

If $F_1 = F_2$ the man is moving a uniform velocity.

121. Worked examples

(g may be taken as 9·8 m/s^2.)

Example 1. *A lift is ascending with an acceleration of* 1·4 *m/s*^{2}. *A man holds a spring-balance from which a parcel of mass* 2 *kg is hung. If the balance is calibrated in kgf, what will be its reading?*

The situation is represented in Fig. 129.

The forces on the parcel are:

Its weight, downwards.
The pull from the string, upwards.

$$\text{The weight} = 2 \times 9{\cdot}8$$
$$= 19{\cdot}6 \text{ N.}$$

Let the pull be T N.

$\therefore$ Resultant force $= T - 19{\cdot}6$ N

FIG. 129.

and this force causes the parcel to accelerate at 1·4 m/s²

$$\therefore \quad \text{Force} = 2 \times 1{\cdot}4$$
$$= 2{\cdot}8 \text{ N}$$
$$\therefore \quad 2{\cdot}8 = T - 19{\cdot}6$$
$$\text{or} \quad T = 22{\cdot}4 \text{ N.}$$

By Newton's Third Law this force must produce an equal and opposite force on the balance.

$$\therefore \quad \text{Force on balance} = 22{\cdot}4 \text{ N}$$
$$= \frac{22{\cdot}4}{9{\cdot}8} \text{ kgf.}$$

∴ Reading on balance is 2·29 kgf.

Example 2. *The effective mass of a certain boat lying in some water is* 300 *kg, and when a man in another boat starts pulling it towards him it accelerates at* 20 *cm/s². If the second man + boat have an effective mass of* 400 *kg, calculate by how much the distance between them is decreased after* 10 *seconds, if a steady force is maintained.*

$$20 \text{ cm/s}^2 = 0{\cdot}2 \text{ m/s}^2.$$
$$\therefore \quad \text{Force on first boat} = M \times a$$
$$= 300 \times 0{\cdot}2$$
$$= 60 \text{ N.}$$
$$\therefore \quad \text{Force on second boat} = 60 \text{ N.}$$
$$\therefore \quad \text{Acceleration of second boat} = \frac{60}{400}$$
$$= 0{\cdot}15 \text{ m/s}^2.$$

After 10 seconds

$$\text{Distance moved by first boat} = \tfrac{1}{2} at^2$$
$$= \tfrac{1}{2} \times 0{\cdot}2 \times 100$$
$$= 10 \text{ m.}$$
$$\text{Distance moved by second boat} = \tfrac{1}{2} \times 0{\cdot}15 \times 100$$
$$= 7{\cdot}5 \text{ m.}$$

∴ Decrease in distance between them is 17·5 m.

Example 3. *A body of mass* 1·5 *kg lies on a horizontal table and is attached, by means of a string passing over a frictionless pulley at the edge of the table, to a body of* *mass 500 g, which hangs freely. As the 1·5 kg body begins to move there is a frictional force acting on it of 0·9 N. Find the initial acceleration of the body.*

The problem is illustrated in Fig. 130.

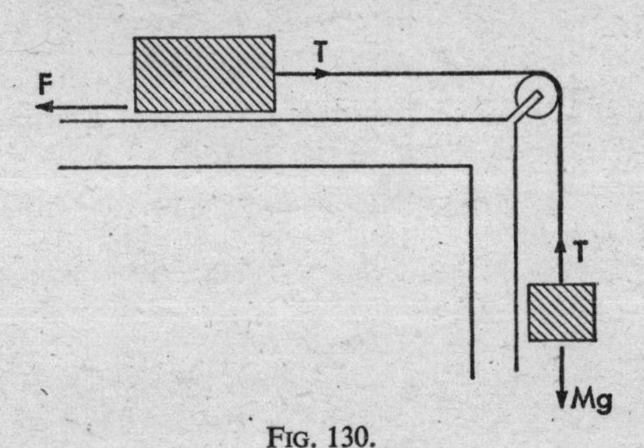

FIG. 130.

If the string remains taut both bodies will accelerate at the same rate.

Let the acceleration be a m/s².

(i) Forces on the 1·5 kg body are:

T Newtons (from the string) forwards.
0·9 Newtons (from friction) backwards.

$$\text{Since } F = Ma$$
$$T - 0{\cdot}9 = 1{\cdot}5a \quad . \quad . \quad . \quad . \quad \text{I}$$

(ii) Forces on the 500-g body are:

T Newtons (from the string) up.
0·5 × 9·8 Newtons (its weight) down.

$$\text{Since } F = Ma$$
$$0{\cdot}5 \times 9{\cdot}8 - T = 0{\cdot}5a \; . \; . \; . \; . \; . \; . \quad \text{II}$$

By adding equations I and II we may eliminate T,

$$\therefore \quad 0{\cdot}5 \times 9{\cdot}8 - 0{\cdot}9 = 0{\cdot}5a + 1{\cdot}5a$$
$$\therefore \quad 4{\cdot}9 - 0{\cdot}9 = 2a$$
$$\therefore \quad 4 = 2a.$$

The acceleration is 2 m/s².

Exercise 17

1. A mass of 5 kg lies on a smooth horizontal table. It is connected by a fine string, passing over a smooth pulley on the edge of the table, to a mass of 2 kg hanging freely. What is the tension in the string and the acceleration of the system?

2. Masses of 25 kg and 24 kg are connected by a fine string passing over a smooth pulley. Find the tension in the string and the force required to support the pulley.

3. Masses of 4 g and 6 g are connected by a string passing over a smooth pulley. With what acceleration does the greater mass descend?

4. A mass of 8 kg, resting on a smooth horizontal table, is connected by a fine string, passing over a smooth pulley on the edge of the table, to 2 kg hanging freely. How far will the 2 kg descend in 4 s?

5. A body of mass 200 g, at rest on a smooth horizontal table, is attached by two threads, passing over pulleys at each end of the table, to masses of 150 g and 130 g hanging freely. The parts of the thread on the table are aligned and parallel to the table. What is the speed of the system 3 s after it is released from rest?

6. Masses of 6 kg and 8 kg hang on opposite sides of a smooth pulley and are connected by a light string. What is the tension in the string and the acceleration with which the system moves?

7. A man in a lift holds a spring balance supporting a 7 kg mass. The lift descends with: (*a*) downward acceleration 4 m/s^2; (*b*) uniform speed 49 m/s; (*c*) upward acceleration 2 m/s^2. What will the balance read in each of these situations?

8. A mass of 6 kg is lowered by a string and has acceleration 2 m/s^2. What is the tension in the string, in N?

9. A man of mass 75 kg is in a lift which is moving upward with an acceleration 2 m/s^2. What force does he exert on the floor of the lift?

CHAPTER XI

BODIES IN COLLISION

122. Types of collision

When a body is dropped onto a floor there are two extreme forms that the collision between the floor and the body may take:

> **Perfectly elastic collision.** In this case the body would bounce back to the height from which it fell.
> **Inelastic collision.** In this case the body would not bounce at all but would remain on the floor.

Usually we have a situation somewhere in between the two. It is an **elastic collision,** but it is not perfectly elastic.

An object bouncing on a floor is a problem dealing with two bodies, one of which is the earth. This is so large that any effect of the object on the earth is small enough to be ignored. In many cases, however, a collision is between bodies of more comparable sizes.

123. Inelastic collisions. Momentum

It is possible to study the effect of two bodies colliding in considerable detail. We are interested in the effect of the collision rather than in that caused by friction, so

FIG. 131.

friction must be kept as low as possible. One method of doing this is to use bodies which are supported on cushions of air, rather like the hovercraft. An inelastic collision can be arranged by making the first body stick to the second body after impact. The arrangement is represented in Fig. 131. The velocities of the bodies before and after the collision are measured, and the sort of results one might obtain are shown in Table III.

Table III

Mass of A	Velocity of A	Mass of B	∴ Mass of $A + B$	Velocity of $A + B$
1 kg	3 m/s	1 kg	2 kg	1·48 m/s
1 kg	3 m/s	2 kg	3 kg	0·98 m/s
1 kg	3 m/s	3 kg	4 kg	0·74 m/s

It will be noticed that both before and after collision the product *Mass of moving body* × *Velocity of body* is approximately equal to 3.

The Momentum of a body is the product of its mass and its velocity.

Momentum = mv.

It is measured in kg m/s.

124. Elastic collisions

Experiments can also be arranged in which the first body does not adhere to the second, but bounces off it.

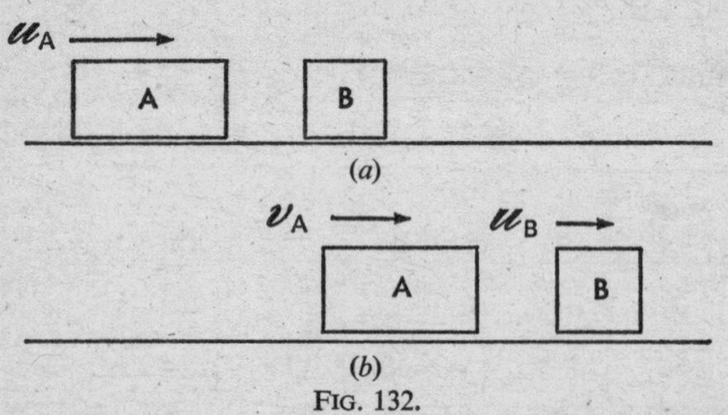

(*a*)

(*b*)

FIG. 132.

It is represented in Fig. 132(*a*) and (*b*). The velocity of the body before impact is represented as u_A, whereas the velocities after impact are shown as v_A and v_B.

Table IV shows a set of idealised results from such an experiment.

Table IV

Mass of A	u_A	v_A	Mass of B	v_B
2 kg	6 m/s	2 m/s	1 kg	8 m/s
2 kg	6 m/s	3 m/s	1 kg	6 m/s
2 kg	6 m/s	3·5 m/s	1 kg	5 m/s

In only one of these results is the collision perfectly elastic. In each case, however,

Momentum of A before impact = Momentum of A after impact + Momentum of B after impact,

e.g. for the first result we have:

Momentum before	*Momentum after*
2×6 kg m/s	$2 \times 2 + 1 \times 8$ kg m/s
12 kg m/s	12 kg m/s

It is left to the reader to check the other results. In every case the final total momentum will be found to equal the original momentum, namely 12 kg m/s.

It is worth studying one further table of idealised results. In Table V we see the possible effect of two bodies

Table V

Mass of A	u_A	v_A	Mass of B	u_B	v_B
1 kg	6 m/s	−2 m/s	2 kg	0	4 m/s
1 kg	6 m/s	−1·5 m/s	1 kg	−4 m/s	3·5 m/s
2 kg	6 m/s	0	1 kg	−3 m/s	9 m/s
2 kg	6 m/s	−2 m/s	1 kg	−6 m/s	10 m/s

which are both moving before they collide as represented in Fig. 133. So far we have considered only movements going from left to right. In Table V the velocities with negative values should be interpreted as being velocities going from right to left.

The last result in the above table represents what might happen when a body of mass 2 kg moving from left to right hits a body of mass 1 kg moving from right to left.

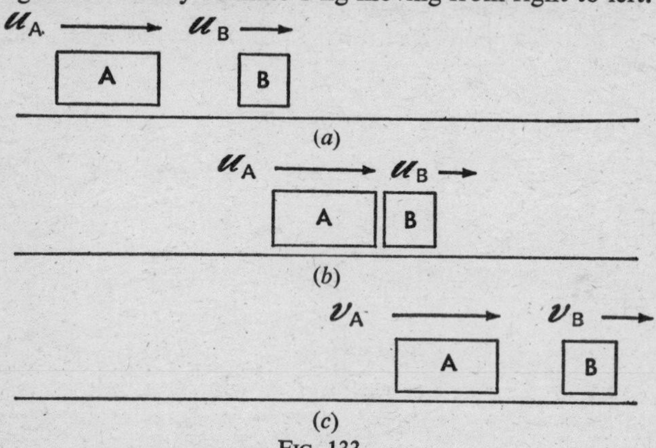

FIG. 133.

Total momentum before impact

$$= \text{Momentum of } A + \text{Momentum of } B$$
$$= 2 \times 6 + 1 \times (-6)$$
$$= 12 - 6$$
$$= 6 \text{ kg m/s.}$$

Total momentum after impact

$$= \text{Momentum of } A + \text{Momentum of } B$$
$$= 2 \times (-2) + 1 \times 10$$
$$= -4 + 10$$
$$= 6 \text{ kg m/s.}$$

It is left to the reader to calculate the total momentum before and after impact for the other results, but when this has been done it will be found that in each case,

Total momentum before impact = Total momentum after impact.

The Law of Conservation of Momentum states that **the total momentum of any system of bodies remains constant provided that they are uninfluenced by external forces.**

125. Worked examples

Example 1. *A body of mass 8 kg moving at 12 m/s collides with another body of mass 16 kg and adheres to it. Calculate the final velocity.*

Since the bodies adhere, both have the same final velocity.

Let the final velocity be v m/s.

Total momentum before impact

$$\begin{aligned} &= \text{momentum of 8 kg mass} \\ &= 8 \times 12 \\ &= 96 \text{ kg m/s.} \end{aligned}$$

Total momentum after impact

$$\begin{aligned} &= (8 + 16) \times v \\ &= 24v \text{ kg m/s.} \end{aligned}$$

By the Law of Conservation of Momentum

$$\begin{aligned} 24v &= 96 \\ \therefore \quad v &= 4. \end{aligned}$$

The final velocity = 4 m/s.

Example 2. *A body of mass 4 kg travelling at 12 m/s collides with a body of mass 8 kg travelling at 2 m/s in the opposite direction. If the second body is caused, by the collision, to return along its original path at a rate of 6 m/s, find the velocity of the first body after impact.*

Let the velocity of the first body after impact be v m/s.

Before impact the total momentum =
Momentum of first body + Momentum of second body

$$\begin{aligned} &= 4 \times 12 + 8 \times (-2) \\ &= 48 - 16 \\ &= 32 \text{ kg m/s.} \end{aligned}$$

After impact the total momentum

$$\begin{aligned} &= 4v + 8 \times 6 \\ &= 4v + 48. \end{aligned}$$

By the Law of Conservation of Momentum

$$4v + 48 = 32$$
$$\therefore \quad 4v = -16$$
$$\therefore \quad v = -4.$$

Thus the first body retraces its original path at a speed of 4 m/s.

Exercise 18

1. What is the momentum of:
   (1) 1 t moving at 45 km/h.
   (2) 1 g moving at 25 cm/s.

2. A ball of mass 0·3 kg moving at 60 cm/s hits a stationary ball of mass 0·40 kg. The first ball rebounds along its approach path with speed 20 cm/s. What is the speed of the second ball?

3. Two blocks of mass 8 kg and 10 kg are connected by a compressed spring. The spring is released and gives the smaller mass an initial speed of 0·6 m/s. What is the initial speed of the other block?

4. A carriage rolling along level track at 10 km/h collides with an identical but stationary carriage. If the carriages are coupled at the moment of collision, what is their speed afterwards?

5. Two trains, of equal mass, shut off their engines immediately prior to a head-on collision, in which they lock together. Their original speeds were 60 km/h and 100 km/h. What is their joint speed immediately after impact?

126. Momentum and Newton's Third Law

Newton's Third Law states that *for every force acting on one body there is an equal and opposite force acting on another body.*

Suppose this force is steady and lasts for a time t seconds.

$$\text{Force} = \text{Rate of change of momentum}$$
$$= \frac{\text{Change in momentum}}{t}.$$

∴ In time t seconds an equal and opposite force on the two bodies will produce an equal and opposite change in momentum. The total momentum of the two bodies, therefore, will be unchanged.

The Law of Conservation of Momentum is thus another way of stating part of Newton's Third Law of Motion.

127. Impulse

In many cases a body does not experience a continuous force, but one which lasts a short but finite time. This is true, for example, of a football being kicked. The foot is in contact with the ball a short time and during that time the ball experiences a force forwards while the foot experiences a retarding force.

If a force F acts on a body for a time t

$F \times t$ is called the **Impulse of the Force.**

It is easy to show that the impulse a body receives is equal to the change in its momentum.

Since "momentum gained" = Force × time, an alternative name for the unit for momentum is a "Newton-second", which is easier to say than a kg m/s.

128. Worked examples

Example 1. *A mass of* 5 *kg rests on a smooth horizontal table. What steady horizontal force applied for* 2 *seconds will cause it to gain a velocity of* 2 *m/s?*

The body has no initial momentum,

$$\begin{aligned} \therefore \quad \text{Gain in momentum} &= 5 \times 2 \text{ kg m/s} \\ &= \text{impulse received} \\ &= F \times 2, \text{ where } F \text{ is the force felt.} \\ \therefore \quad F &= 5 \text{ Newtons.} \end{aligned}$$

Example 2. *A man pushes a truck of mass* 2500 *kg along horizontal rails with a steady force of* 300 *N. Resistance* *amounts to* 40 *N per* 1000 *kg. Find the velocity of the truck at the end of* 30 *s and the distance moved.*

$$\text{Total Resistive Force} = 2{\cdot}5 \times 40$$
$$= 100 \text{ N.}$$
$$\therefore \quad \text{Resultant force on truck} = 300 - 100$$
$$= 200 \text{ N.}$$
$$\therefore \quad \text{Impulse received by truck} = 200 \times 30$$
$$= 6000 \text{ kg m/s.}$$

But impulse received = increase in momentum.

$$\therefore \quad mv = 6000$$
$$\text{or} \quad 2500v = 6000$$
$$\therefore \quad v = 2{\cdot}4 \text{ m/s.}$$

To find the distance covered use

$$s = \frac{u + v}{2} \times t$$
$$= \frac{2{\cdot}4}{2} \times 30 \text{ (since } u = 0)$$
$$= 36 \text{ m.}$$

Exercise 19

1. A lorry of mass 1·5 t is moving at 36 km/h. What force, in kN, is required to bring it to rest in 12 s?
2. A mass of 5·8 kg is moving at 25 m/s. (*a*) What is its momentum? (*b*) If the mass, originally at rest, has been moving for 10 s, what average force has been acting?
3. A mass of 75 kg moving at 16 m/s is brought to rest in 2 s. What is the magnitude of the resisting force?
4. What is the force which, acting on a mass of 60 kg for 12 s, will produce a change of 20 m/s in the speed?
5. A force of 6 N acts for 8 s upon a mass of 24 kg, initially at rest. What is the change in momentum and the final velocity?
6. The speed of a 12 kg mass is 50 m/s. In 4 s a constant force reduces the speed to 30 m/s. What was the magnitude of the force?

CHAPTER XII

GRAVITATION, MASS AND CIRCULAR MOTION

129. Newton's Law of Gravitation

So far we have deduced that, since bodies accelerate towards the earth when released, the earth must be attracting them. We now consider the nature of this attraction. Newton examined a lot of data concerned with the motion of the planets, and decided that the movements could be explained in terms of forces of attraction between them. His result is summed up in the following equation.

$F = \frac{Gm_1m_2}{r^2}$ where F is the force of attraction between bodies of mass m_1 and m_2,
r is the distance between them,
G is a constant.

In the m.k.s. system of units, which is used in this book, the value of G is about $6{\cdot}7 \times 10^{-11}$, so it is easy to see that the force of attraction is very small unless the masses are very large. Nevertheless the force is sufficient to pull a plumb-line out of the vertical plane when it is hanging near a very large mountain. The actual vertical direction is taken from astronomical observations. Very accurate experiments in laboratories can show up the attraction between two lead spheres, and such experiments have been used to determine the value of G.

130. Gravitational force and weight

Consider a body of mass m on the surface of the earth, which we take to have mass M and radius R.

Force of attraction between earth and body

$$= \frac{GMm}{R^2}.$$

This is the force caused by gravitational attraction, and we usually refer to it as the weight.

The complete expression for the weight is slightly more complicated, because the mass m not only feels an attraction towards the earth, but also small attractions towards the moon and the sun. As the moon changes its position relative to the earth the strength of this contribution varies, so the weight is continually changing by a *very* tiny amount. Small as these extra forces are, it is they which are responsible for the production of tides. A quick glance at any tidal chart will convince the reader that there is a close link between the cycle of the tides and the cycle of the moon.

131. Gravitational force and gravitational acceleration

The force between the body and the earth is

$$F = \frac{GMm}{R^2}.$$

$$\therefore \text{ Acceleration } = \frac{F}{m} = \frac{GMm}{mR^2} = \frac{GM}{R^2}$$

$$\text{i.e.} \quad g = \frac{GM}{R^2}.$$

It will be seen from this that, at a given spot on the earth's surface, the **acceleration due to gravity is independent of the mass of the body,** and hence is independent of its weight. This result was pronounced by Galileo about 100 years before Newton.

132. Practical methods for comparing masses

In § 110 a fundamental method for comparing the masses of two bodies was described. It involved pulling the two bodies with the same force and measuring the accelerations that each gained. As a practical method of comparing masses this had many disadvantages, and it is much more usual to make use of the fact that, at the same place, bodies with equal masses will feel equal forces of attraction to the earth. In other words they have the same weight at that place,

i.e. $$M_1g = M_2g$$
$\therefore$ $$M_1 = M_2.$$

133. Comparing masses with a beam balance

For this method two equal pans are suspended from a light pivoted beam at equal distances from the fulcrum. When the pans are empty the beam should be horizontal. The body which is of unknown mass is then added to the left-hand pan, and standard known masses are added to the right-hand pan until the beam is again horizontal.

FIG. 134.

Application of the principle of moments shows that the forces acting on each side of the beam must be equal, and, since both of the forces are the result of gravitational attraction acting on the masses A and B (Fig. 134), it follows that these masses must both be equal.

This method of comparing masses would work equally well wherever the experiment was performed, so long as there was a gravitational attraction to act on both the masses.

134. Comparing masses with a spring balance

If the body of unknown mass is hung from a spring balance the force of gravity will act on it and pull it downwards. The position of the pointer should be noted.

The body is then removed and replaced by combinations of known masses, until the pointer again takes up the same position. This shows that the force of gravitational attraction in both cases is the same, and therefore, *since both experiments are performed at the same place* and almost at the same time, it is reasonable to suppose that the masses must be equal in both cases.

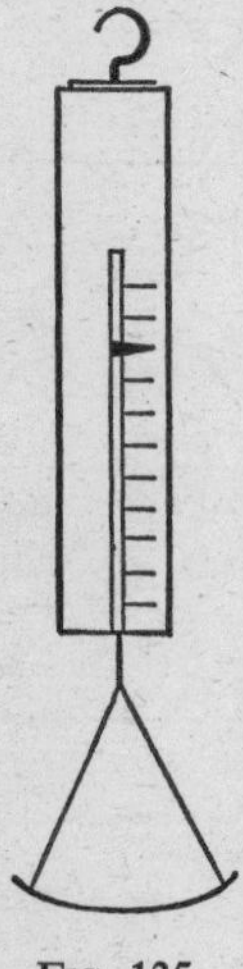

Fig. 135.

Calibrated "scales"

A spring balance could be calibrated by adding different standard masses and noting the extension produced by the earth's pull on each. Thus one mark on the scale would correspond to the extension of the spring caused by the earth pulling on 1 kg, and another mark would correspond to that caused by the earth pulling on 2 kg. If another object caused an extension exactly half-way between, we would be tempted to assume that its mass must be 1·5 kg. This would be fair, however, *only if it were used in the place where it had been calibrated.* If such an instrument, which had been calibrated on the earth, were used on the moon, a mass of 6 kg on the moon would appear to have a mass of only 1 kg, because the earth attracts about 6 times as strongly as the moon. An object, when released, accelerates 6 times as fast on the earth as it does on the moon. Most of the scales used by shop-keepers use some form of spring system. If the scales are moved from one part of the country to another, 1 kg of potatoes will give a slightly different reading in the two places. The change from John O'Groats to Lands End is only 0·03 per cent., however, so it is not of practical significance. Even if it were a bigger change it would be easy to recalibrate the scales by checking the readings when standard masses were placed on them.

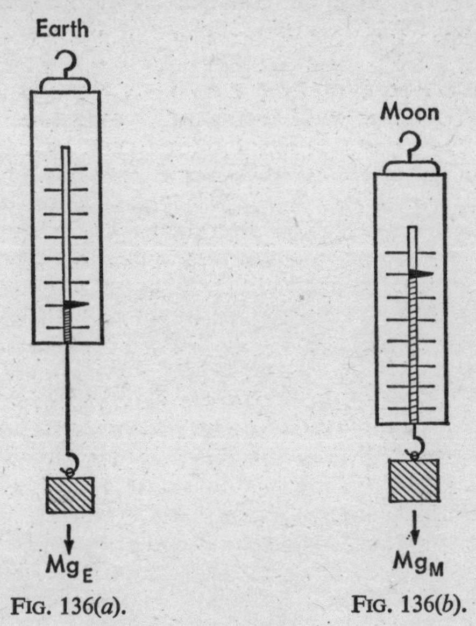

FIG. 136(*a*). FIG. 136(*b*).

135. Summary of mass and weight

Mass

(*a*) It measures the inertia of a body.

(*b*) It does not change from place to place.

(*c*) Two masses can be compared by measuring the accelerations produced when both are subjected to the same force.

(*d*) A better method is to compare the gravitational attraction on the two bodies at the same place at the same time. This can be done with a beam balance.

Weight

(*a*) It measures the force of gravitational attraction between a body and the larger body on which it rests.

(*b*) It changes from place to place.

(*c*) It can be measured by the extension of a spring balance.

The weight of a body varies only slightly at different places on the earth, so a comparison of weights is effectively a comparison of the masses of two objects.

136. Movement in a circle. Centripetal force

Momentum, like velocity (see Chapter VIII), is a vector. This means it has both magnitude and direction. Usually when we talk about changes in momentum we are thinking about changes in magnitude, but a change in direction without change in magnitude also represents a change in momentum. The simplest example of this is a body being whirled round in a circle. The speed may be constant, but the velocity is constantly changing direction. So also is the momentum.

Application of Newton's First Law shows that there must be a force. In the absence of any force the body would move in a straight line, as in Fig. 137(*a*). If it

FIG. 137(*a*).

keeps changing its direction, as in Fig. 137(*b*), we deduce that there is a force, F, which must be acting perpendicular to the direction of travel.

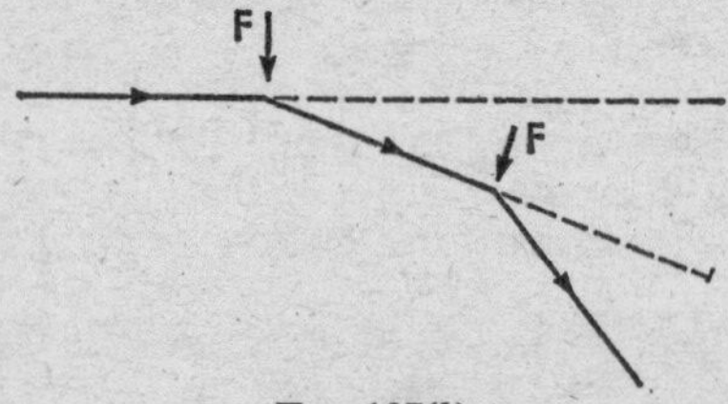

FIG. 137(*b*).

When a body moves in the arc of a circle its direction is continually changing, and this must be caused by a steady force. This force is called the centripetal force, because it is directed towards the centre of the circle.

When a body is being whirled round on the end of a string, the tension in the string provides the centripetal force necessary to keep the body moving in a circle. If the string is released the force is removed and the body

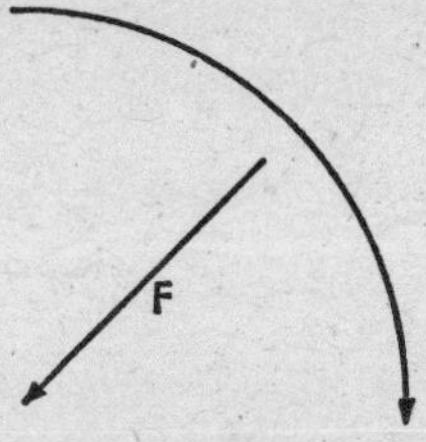

FIG. 137(*c*).

obeys Newton's First Law by travelling along the tangent to the circle, i.e. by continuing in the direction in which it was travelling when the string was released. This is illustrated in Figs. 138(*a*) and (*b*).

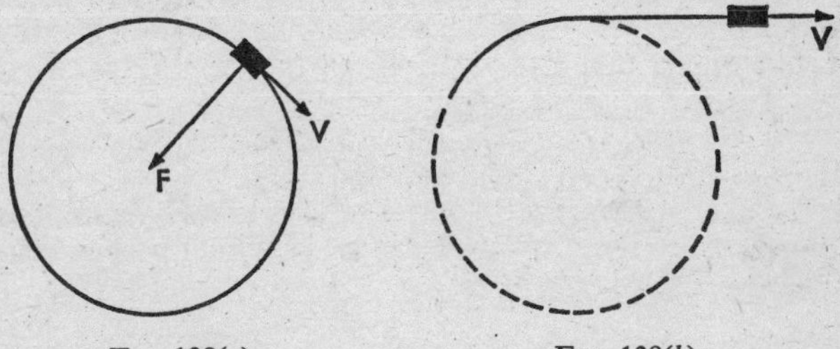

FIG. 138(*a*). FIG. 138(*b*).

The Centripetal Force is the force necessary to keep a body moving in a circle.

It can be shown that for a body moving in a circle,

Centripetal Force $= \dfrac{mv^2}{r}$ Newtons, where m kg is the mass,
v m/s is the speed,
r m is the radius of the circle.

137. Movement of satellites

Consider a satellite of mass m moving at v m/s parallel to the surface of the earth at a height h m above it. There is only one force acting on the satellite, that of gravity.

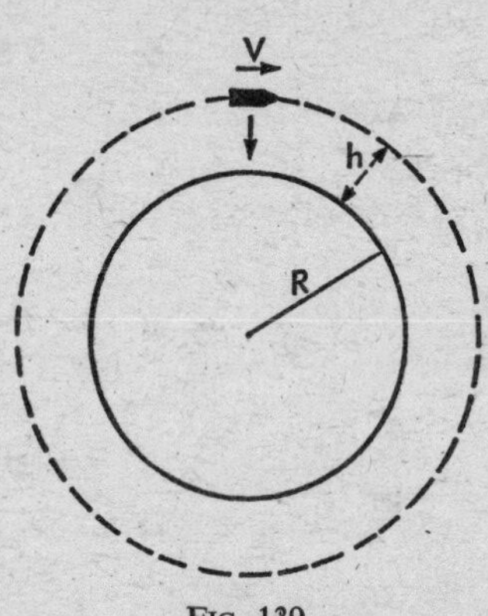

FIG. 139.

Force of gravity $= \dfrac{GMm}{(R+h)^2}$ where M = mass of earth.

R = earth's radius.

The force needed to keep the satellite moving in a circle of radius $(R + h)$ is

$$\frac{mv^2}{(R+h)}.$$

This force is perpendicular to the direction of flight, and therefore the satellite changes direction.

If $\dfrac{mv^2}{R+h} = \dfrac{GMm}{(R+h)^2}$ the satellite will move in a circle of radius $(R + h)$.

138. Worked examples

Example. *A satellite circles the earth* 300 *km above the earth's surface. Find the time it takes to complete one* *orbit. Take the radius of the earth as* 6400 *km and its mass as* 6×10^{24} *kg. Newton's constant of Gravitation in m.k.s. units may be taken as* $6{\cdot}7 \times 10^{-11}$.

Using the symbols of § 136,

$$\text{Gravitational attraction} = \frac{GMm}{(R+h)^2}.$$

$$\text{Centripetal force needed} = \frac{mv^2}{(R+h)}.$$

$\therefore$ if it moves in an orbit of total radius $(R + h)$

$$\frac{mv^2}{(R+h)} = \frac{GMm}{(R+h)^2}.$$

$$\therefore \qquad v^2 = \frac{GM}{R+h}.$$

The distance travelled in one orbit $= 2\pi(R + h)$.

$$\therefore \quad \text{Time taken} = \frac{2\pi(R+h)}{v}$$

$$= 2\pi \frac{(R+h)^{\frac{3}{2}}}{\sqrt{GM}}$$

$$R + h = 6400 + 300 \text{ km}$$

$$= 6{\cdot}7 \times 10^6 \text{ m.}$$

$$\therefore \quad \text{Time taken} = 2\pi \frac{\sqrt{(6{\cdot}7 \times 10^6)^3}}{\sqrt{6{\cdot}7 \times 10^{-11} \times 6 \times 10^{24}}}$$

$$= 2\pi \times 6{\cdot}7 \times 10^2 \times \frac{10}{6}$$

$$= 28{,}680$$

$$= 4{\cdot}78 \text{ h approx.}$$

Example 2. *A car of mass 700 kg travels round a bend, which is part of a circle of radius* 100 *m, at* 72 *m km/h. If the bend is banked at such an angle that the car has no tendency to move either up or down the slope, find the angle of banking.* 72 km/h = 20 m/s.

The situation is shown in Fig. 140. Usually there would be friction, but, since there is no tendency to move up or down the slope, the only forces on the car are:

Mg the weight;
R the reaction of the slope.

Since the weight can have no horizontal component, the centripetal force must be provided by R.

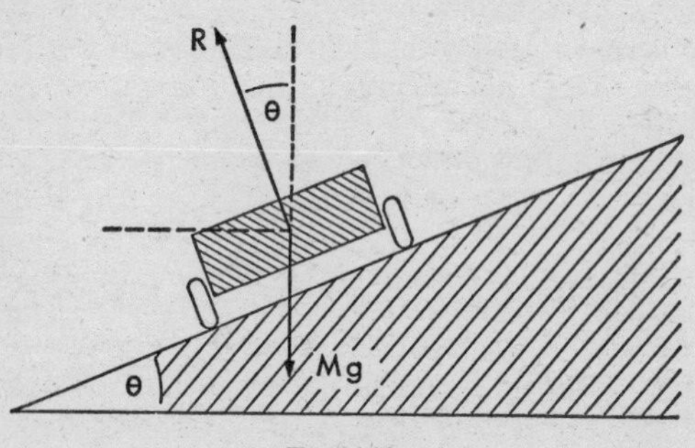

FIG. 140.

Resolving R vertically and horizontally we have:

$$R\cos\theta = Mg \text{ (no vertical movement)}$$

$$R\sin\theta = \frac{Mv^2}{r} \text{ (centripetal force)}$$

$$\therefore \quad \frac{R\sin\theta}{R\cos\theta} = \frac{Mv^2}{rMg}$$

$$\therefore \quad \tan\theta = \frac{v^2}{rg}$$

$$= \frac{20^2}{100 \times 9{\cdot}8}$$

$$= 0{\cdot}408$$

$$\therefore \quad \theta = 22^\circ\,25' \text{ approx.}$$

N.B. (*a*) If v increased, there would not be sufficient force to keep the car moving in the circular arc unless some friction acted down the plane.

(*b*) If v were decreased the car would slide down the slope unless a frictional force acted up the slope.

139. "Weightlessness" and "free fall"

If a man in a lift, which is accelerating downwards with the acceleration due to gravity, releases a ball, it cannot accelerate at a greater rate than g and thus the man will find that the ball does not fall relative to his hand at all. Both the man's hand and the ball are accelerating at the same rate and therefore they remain together.

The same situation is true in an orbiting space-craft. The space-craft feels a gravitational attraction, and it is this which stops it from travelling along a straight line, and makes it move in a circle. The space-craft accelerates towards the earth, and if the earth were flat it would approach it. Since the earth is curved the acceleration is just sufficient to stop the space-craft moving farther away. Any object in the space-ship feels a force from the earth, and experiences an exactly similar acceleration. Thus when a body in an orbiting space-craft is released it floats relative to the space-craft. We say that it is in "free fall".

Exercise 20

$$\left(g = 9{\cdot}8 \text{ m/s}^2;\ \pi = \frac{22}{7}.\right)$$

1. A satellite rotates in a circle around its parent planet. The period of rotation is 76 min and its speed is 220 m/s. How far are the planet and satellite apart?

2. A piece of lead, of mass 1·4 kg, is swung round at the end of a string which is 0·7 m long. It travels in a horizontal circle with a period of 2 s. What is the tension in the string?

3. What force is acting on a man of 75 kg, who is sitting in a train, travelling at 90 km/h, on a circular section of track, if the radius of the section is 100 m?

4. A car takes a circular bend, of radius 30 m, at

45 km/h. What is the least coefficient of friction that can prevent side-slipping?

5. A van is steered round a horizontal curve of radius 25 m. The van is 2 m wide and its centre of gravity is 1·25 m above the ground. What is the highest speed that the van can corner at without toppling over?

CHAPTER XIII

WORK, ENERGY AND POWER

140. Work

When a body is moved by the action of a force the force is said to do work.

Work in this sense is one of the most frequent occurrences in our existence. When a man lifts a weight to a height from the ground the muscular force which he exerts does work. If it is a large weight he will need to use a large force. He will do a large amount of work. If he lifts the weight twice as high, we would expect him to have done twice as much work.

The work, therefore, is connected with the size of the force, and the distance through which the force moves.

We define work in the following way:

The **work done by a force** = The size of the **force** × **the distance moved** by the force *in the direction of the force.*

Some people omit the words in italics, but the following example will illustrate the necessity of their inclusion.

Worked example

A body of mass M kg slides down a perfectly smooth plane, inclined at an angle θ to the horizontal, under the action of gravity. Find the work done by gravity if the slope is b metres long.

The situation is represented in Fig. 141. The forces acting on the body are:

(i) Gravitational attraction, vertically down,

$$= Mg \text{ Newtons.}$$

(ii) The reaction of the inclined plane, R, perpendicular to the plane.

The gravitational force of the body may be resolved into components parallel to and perpendicular to the plane.

Component parallel to plane $= Mg \sin \theta$.
Component perpendicular to plane $= Mg \cos \theta$.

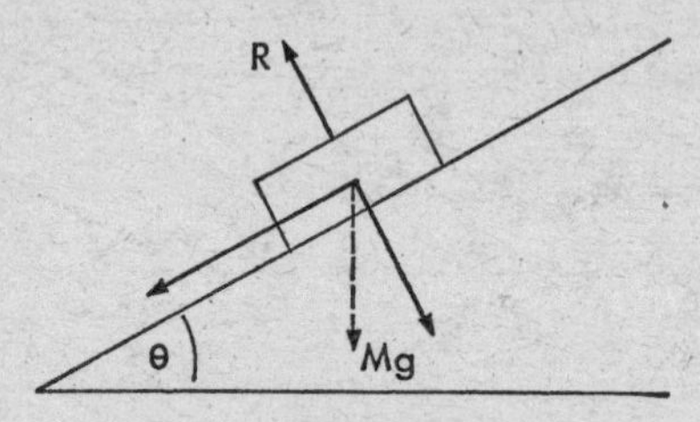

FIG. 141.

Since there is no movement perpendicular to the plane,

$R = Mg \cos \theta$.
Work done $= Mg \sin \theta \times b$;
but $\sin \theta \times b = y$, where y is the vertical distance moved, i.e. the distance moved in the direction of the gravitational force.

Thus work done by gravitational force

= gravitational force $\times\ y$
= gravitational force $\times$ distance moved in direction of gravitational force.

141. The Joule

The unit of work is the Joule.

One Joule of work is done when a force of one Newton moves its point of application by one metre.

A force of 1 N moving 1 m does 1 J.
A force of F N moving 1 m does F J.
A force of F N moving s m does $F \times s$ J.

Work done = $F \times s$ Joules.

142. Work done in lifting objects

If an object has a mass of m kg it is attracted to the earth by a force of mg Newtons, where g is the acceleration due to gravity.

To lift the object we need a force of mg Newtons.

In lifting it through h metres

$$\text{Work done} = mgh \text{ Joules.}$$

Work is done only while it is being lifted. If it is then kept in a raised position, e.g. by resting it on a table, no further work is done.

143. Work done in holding objects

To tell a person who is holding a heavy object at arm's length that he is doing no work is to invite reprisals! And yet as he holds it there, although there is undoubtedly a force, *the force is not moving its point of application*, and so apparently no work is being done. The answer to this apparent paradox comes from examining the person's muscles as he stands there. Far from being stationary these muscles are continually contracting and relaxing, and the effect is the same as if the person were continually lowering the object and lifting it up again.

He is, therefore, doing work

= muscular force $\times$ distance moved; but the distance is the sum of lots of small distances.

144. Work done against friction. Example

If a body is moving at a uniform velocity over a surface with friction the applied force forwards must equal the frictional force.

A body is pulled along a rough surface at a uniform velocity by a force of 500 *N inclined at* 36° *to the horizontal. Calculate the work done in moving the body* 7 *m forward.*

The situation is represented in Fig. 142. The forces on the body in the horizontal direction are 500 cos 36

= 400 N forwards

and F_F backwards, where F_F is the frictional force. Since there is no acceleration the two forces are equal. The work done

= work done against friction
= 400 × 7
= 2800 Joules.

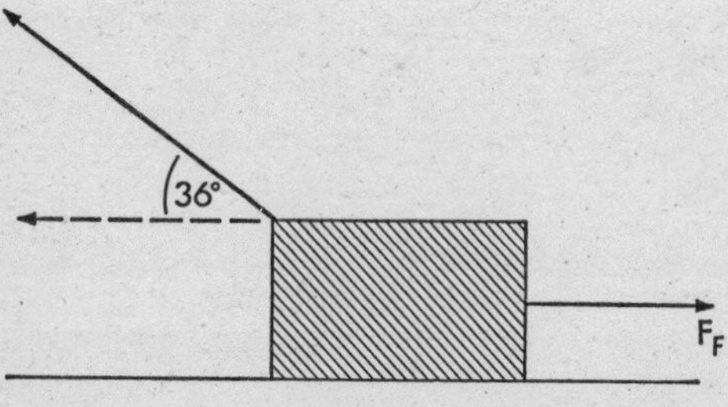

FIG. 142.

N.B. The vertical component of the applied force, which is 500 sin 36, does no work because there is no movement in the vertical direction.

145. Energy

When a man uses muscular force to lift a weight we say that he has used up some energy. The more work he does the more energy he uses. He can replace the energy by eating some more food, and it is worth noting that a heavy manual worker needs to eat more than, say, an office worker.

Energy is the capacity for doing work

It is measured in the same units as the work it can do, i.e. it is measured in Joules.

There are many forms of energy. We shall mention just a few.

146. Kinetic energy

This is the energy a body has by virtue of its motion. As it is slowed down it exerts a force on the retarding agent, and this force can do work.

Consider a body that is slowed down from v m/s to rest in a distance s m.

Using $v^2 - u^2 = 2\,as$ we get, in this case,

$$a = -\frac{v^2}{2s}.$$

The negative sign indicates that the body is being retarded.

$$\therefore \quad \text{the force on the body} = -\frac{mv^2}{2s},$$

i.e. a retarding force.

$$\therefore \text{ force on retarding agent} = \frac{mv^2}{2s}$$

(Newton's Third Law).

$$\text{Work done} = \text{force} \times \text{distance moved}$$

$$= \frac{mv^2}{2s} \times s = \frac{1}{2}mv^2.$$

The kinetic energy of a body

$$= \tfrac{1}{2}\boldsymbol{mv^2}.$$

147. Potential energy

If an object is lifted through a height h

Work done $= mgh$ J (*see* § 142).

If it is then released and falls through the same height h metres, its final velocity, v m/s, is given by

$$v^2 - u^2 = 2as$$

or, in this case, since $u = 0$ and $f = g$,

$$v^2 = 2gh$$

∴ its kinetic energy

$$\begin{aligned} &= \tfrac{1}{2}mv^2 \\ &= \tfrac{1}{2}m2gh \\ &= mgh \text{ Joules.} \end{aligned}$$

Thus the work done in lifting the body reappears as kinetic energy when the body is released and allowed to fall freely. We say that the body has effectively stored the energy by being in the raised position. This stored energy is called potential energy. In this case we have considered gravitational potential energy.

For a body of mass *m* kg lifted through a height *h* metres the

gravitational potential energy = mgh Joules.

There are many other forms of potential energy. We get energy stored in the spring of a clock when we wind it up. We can get energy associated with electricity or with atomic and nuclear forces.

148. Conservation of energy

In § 144 we talked of a man using up energy. His energy was converted into the work done in lifting the weight. It was converted into potential energy. Scientists now have considerable confidence in the belief that:

Energy can neither be created nor destroyed.

In every case where energy may have been destroyed it is believed that it simply changes form. In a great many cases the end product is **heat energy.** This is what **is produced when we do work against friction.** It is for this reason that brake linings get so hot.

It is instructive to consider some of the sources of energy:

To turn the pages of this book you needed energy.

This energy was supplied by the effect of certain chemical reactions in your body, acting on your food.

The food originally absorbed energy in the process of growing.

The plant which supplied the food may have been helped to grow by an electric heater.

The electrical energy may have come from a generator worked by a water turbine.

The water driving the turbine had kinetic energy.

It gained its kinetic energy by conversion of gravitational potential energy, which may have been quite considerable if the water was stored in a high dam.

The water reached the dam as rain. Earlier it had received heat energy from the sun to enable it to evaporate from the oceans.

The sun's heat energy is provided by nuclear reactions on the surface of the sun.

All of this is in the broader realms of physics rather than mechanics. It is worth mentioning, in passing, an extension to the Law of Conservation of Energy. It has been found that when energy is released in reactions a very, very tiny amount of mass is lost. This is in accordance with the famous equation:

$E = mc^2$ where E is the energy produced,
m is the mass,
c is the velocity of light.

149. Worked examples

Example 1. *A bullet of mass* 5 g *has a speed of* 500 m/s. *What is its kinetic energy? How far will it penetrate a fixed block if the latter offers a constant resistance of* 1 kN *to the motion of the bullet?*

$$\text{K.E.} = \tfrac{1}{2}mv^2$$

$$\therefore \quad \text{K.E. of bullet} = \frac{1}{2}\,\frac{5}{1000}\,(500)^2 = \frac{2500}{4} = 625 \text{ J.}$$

N.B. m must be in kg.

Let s be the distance of penetration.

Work done by bullet against resistance of block

$$= 1000s \text{ J}$$

and this energy came from the K.E.

$$\therefore \quad 625 = 1000s$$

$$\therefore \quad s = 0{\cdot}625 \text{ m.}$$

Example 2. *A car of mass* 1 *tonne and travelling at* 90 *km/h is brought to rest in* 50 *m by applying the brakes. What was the frictional force exerted on the tyres by the road, assuming that it was uniform?*

The frictional force causes the car to lose K.E. and it acts through 50 m.

Using $Fs = \frac{1}{2}mv^2$,

where $F =$ the frictional force

$m = 1000$ kg

$$v = \frac{90\,000}{60 \times 60} = 25 \text{ m/s}.$$

$$\therefore \quad F \times 50 = \frac{1000 \times 25 \times 25}{2}.$$

$$\therefore \quad F = 250 \times 25 = 6250 \text{ N}.$$

Example 3. *A waggon of mass* 12 *tonnes runs freely down an incline of* 400 *m the slope of which is* 1 *in* 100. *Find the K.E. of the waggon at the bottom of the incline, neglecting friction. Find also the velocity at the end of the run.*

A slope of 1 in 100 means that the waggon drops 1 m for every 100 m travelled along the track.

∴ It drops a total of 4 m.

$$\therefore \quad \text{Potential energy lost} = mgh = 12\,000 \times 9{\cdot}8 \times 4$$

and this is converted to K.E.

$$\therefore \quad \text{K.E.} = 12\,000 \times 9{\cdot}8 \times 4 = 470 \text{ kJ}.$$

Since $\text{K.E.} = \frac{1}{2}mv^2$

$$\tfrac{1}{2} \times 12\,000 \times v^2 = 12\,000 \times 9{\cdot}8 \times 4$$

$$\therefore \quad v^2 = 9{\cdot}8 \times 8$$

$$\therefore \quad v = 8{\cdot}85 \text{ m/s}.$$

Exercise 21

1. Find the kinetic energy of:

(1) a mass of 20 kg at speed 6 m/s;
(2) a mass of 2 t moving at speed 18 km/h.

2. A motor cyclist with his machine weighs 150 kgf and they are travelling at 36 km/h. What is their kinetic energy?

3. What work must be done on a car of mass 400 kg to give it a speed of 20 m/s?

4. A lorry of mass 2 t is moving at 18 km/h. What force will bring it to rest in 20 m?

5. A bullet of mass 95 g has a speed of 400 m/s. In what distance will it come to rest if it is moving against a resistance of 400 N?

6. A boy of mass 60 kg starts to move along a horizontal slide with a speed of 3 m/s and comes to rest after 20 m. What is the average resistance?

7. A car of mass 400 t and travelling on a level road at 36 km/h is brought to rest in 20 m. What is the average resistance?

8. How much K.E. is acquired by a mass of 2 kg falling from rest through 50 m? How much K.E. is acquired falling through the next 50 m?

9. A train of mass 400 t is running on the level at 36 km/h. The engine is cut off, reducing the speed to 18 km/h. What is the average resistive force acting?

10. What is the K.E. of a bullet, mass 10 g, travelling at 500 m/s? What percentage of its energy will it have lost after passing through a block of wood, 300 mm thick, which offers an average resistance of 400 N?

11. A stone weighing 2·4 k falls from the top of a building 15 m high, and penetrates 6·7 m into the ground before coming to rest. What is the average resistance of the ground?

12. A ship whose mass is 10 Mg, moving with velocity 1·4 m/s, shuts its engine off. It travels 40 m before coming to rest. What is the average resistance of the water to the ship?

150. Power. The Watt

The engineer uses the word power in a strict technical sense.

Power = the rate of doing work

i.e. Power = Work done/second

or Power = Energy converted/second.

In both cases it is measured in the same units, J/s or Watts.

1 Watt = 1 Joule/second.

The usual prefixes are used to indicate multiples of a watt. Thus:

1 2-kW electric fire provides 2000 Joules/second in the form of heat energy. A 2-GW power station can transmit 2 000 000 000 Joules/second in the form of electrical energy. Parts of this could provide the power for driving an escalator, a water pump, or many other devices.

151. Power produced by a moving force

$$\text{Power} = \text{Work done/second.}$$

But $\text{Work done} = F \times s$

$$\therefore \quad \text{Power} = \frac{F \times s}{t}$$

$$\frac{s}{t} = v, \text{ the velocity}$$

$$\therefore \quad \textbf{Power} = \boldsymbol{F \times v}.$$

152. Horse Power

A much older unit of power, which the student may meet in older pieces of technical literature, is the Horse Power. This unit was devised by the great engineer James Watt, and it was his estimate of the rate of which a good horse could work for a few hours, although he deliberately fixed it at a high amount.

1 H.P. = 550 ft lbf/s
Now 1 lb = 0·45 kg
∴ 1 lbf = 0·45 kgf
1 kgf = 9·81 N
1 ft = 0·305 m
∴ 1 H.P. = 550 × 0·305 × 0·45 × 9·81 Watts
= 746 W.

153. Worked example

What power is needed to draw a train of mass of 200 *tonnes up an incline of* 1 *in* 200 *at a speed of* 18 *km/h, when the frictional resistance to motion is* 50 *N/tonne?*

Work is done

(1) in raising the train against gravity;
(2) in overcoming frictional resistance.

Now $$18\ \text{km/h} = 5\ \text{m/s}$$

$\therefore$ Train is being raised at a rate of $\frac{5}{200}$ m/s.

Gravitational force on the train

$$= 200\,000 \times 9{\cdot}8\ \text{N}$$
$$= 3860\ \text{kN}.$$

$\therefore$ (1) Power needed to raise train

$$= F \times v$$
$$= 3860 \times \frac{5}{200}$$
$$= 49\,000\ \text{W}$$
$$= 49\ \text{kW}.$$

(2) Power needed against friction

$$= F \times v$$
$$= 200 \times 50 \times 5$$
$$= 50\ \text{kW}.$$

$\therefore$ Total Power needed

$$= 49 + 50 = 99\ \text{kW}.$$

Note

$$\text{Power} = F \times v.$$

If the force is expressed in kN and the velocity in m/s the power will be in kW.

Exercise 22

1. What is the power of an engine which pulls a train of mass 150 t up a slope of 1 in 100 at 40 km/h, assuming that there is no resistance?

2. A load of 1·2 kN is raised from a depth of 200 m, in 4 min, to ground level. What was the power developed?

3. A pump raises 120 t of water an hour through a height of 18 m. Neglecting frictional resistance, what is the power of the engine?

4. A gang of men raised 2000 concrete blocks in an hour. Each block was 3 kg and was raised through an average height of 7 m. What was the power developed?

5. A car was travelling at 40 km/h, on the level. If the engine developed 7 kW, what was the resistance to the car's motion?

6. The resistance to a liner moving at high speed is 220 tf. What power is necessary to drive the ship at 25 knots (1 knot = 0·51 m/s)?

7. A train and locomotive of mass 270 t is moving at 50 km/h and the resistance overcome is 250 N; what is the power of the engine?

8. Find the power required to make a pump raise 1000 kg of water each minute through a height of 15 m, if the water flows off with a speed of 4 m/s.

9. A man of 75 kg runs up his stairs, which are 4 m high, in 5 s. What power did he develop?

10. A car of mass 1200 kg accelerated through 30 km/h in 4 s on a level road. If the frictional resistance is 50 N, what is the average power used?

11. A ship requires 25·5 MW to drive it at 30 knots. What is the resistance to the motion of the ship?

12. A cyclist is riding at 18 km/h against resistances amounting to 13·5 N. What power is he producing?

CHAPTER XIV

MACHINES

In Chapter II a machine was described as a device which enables a small effort to move a large load. The lever, which is one of the best known and most useful machines, was dealt with at some length, because the physical principles are very important. In this chapter we shall examine many other types of machine and see that in some cases the definition needs to be extended.

The general principle underlying all machines is that the work applied to the machine should be employed in a more advantageous manner.

154. Load and effort

The effort is the force applied to the machine.

The machine operates on the **load.** In the simplest example the machine lifts the load, and thus the load is the force applied by the machine.

155. Mechanical advantage

$$\textbf{Mechanical advantage} = \frac{\textbf{Force applied } \textbf{\textit{by}} \textbf{ the machine}}{\textbf{Force applied } \textbf{\textit{to}} \textbf{ the machine}}$$

or

$$\textbf{M.A.} = \frac{\textbf{Load}}{\textbf{Effort}}.$$

The amount of friction that is present will affect the value for the mechanical advantage, but even in the presence of friction we would expect M.A. to be > 1. For a few purposes, however, it is useful to devise a machine which would have a M.A. which is < 1 even if there were no friction.

156. Machines and work

Although it is possible for a small effort to lift a large load it is not possible for a small amount of work put into a machine to result in a lot of work being done. Such a machine would be against the Law of Conservation of Energy. Many ingenious attempts have been made to create such a machine, and at one time in the history of science as much effort was wasted in trying to make such a machine as was wasted in trying to find a chemical reaction in which gold was made from base metals.

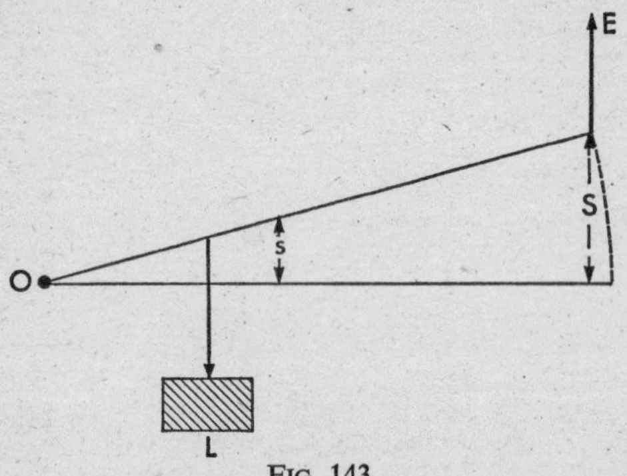

FIG. 143.

Consider now a simple machine in the form of a lever, by means of which an effort E lifts a load L. Let the effort move a distance S and the load move a distance s. It is illustrated in Fig. 143.

Work done *on* machine by effort

$$= E \times S = \textit{Input work.}$$

Work done *by* machine on load

$$= L \times s = \textit{Output work.}$$

As stated above the output work cannot be greater than the input work, and usually

Input = Output + Work done against friction, etc.

157. Efficiency

If the work done by a machine were equal to the work put into it, we should have a perfectly efficient machine.

Since some energy (work) is used against frictional forces it is of interest for us to know what fraction of the work done on the machine is gainfully used.

$$\textbf{Efficiency} = \frac{\textbf{Useful work done by machine}}{\textbf{Total work done on machine}} = \frac{\textbf{Output}}{\textbf{Input}}.$$

This ratio is always less than unity, and the closer to unity it approaches the more efficient the machine. Often a mere spot of oil will greatly increase the efficiency.

158. Velocity ratio

For a perfect machine

$$\text{Output} = \text{Input}$$
$$L \times s = E \times S$$
$$\therefore \quad \frac{L}{E} = \frac{S}{s}.$$

This ratio $\frac{S}{s} = \frac{\textbf{Distance moved by effort}}{\textbf{Distance moved by load}}$ is known as the **velocity ratio** for the machine.

The name was originally chosen because machines involve movement, and

$$\frac{S}{s} = \frac{\text{Distance moved by effort in given time}}{\text{Distance moved by load in same time}}$$
$$= \frac{\text{Velocity of effort}}{\text{Velocity of load}}.$$

The velocity ratio (V.R.) of a machine depends purely on geometrical considerations and can always be calculated. *The velocity ratio is independent of external factors such as friction.*

For any machine

$$\text{Efficiency} = \frac{\text{Output}}{\text{Input}} = \frac{L \times s}{E \times S} = \frac{L}{E} \times \frac{s}{S}$$
$$= \text{M.A.} \div \text{V.R.}$$

i.e.
$$\text{Efficiency} = \frac{\text{Mechanical advantage}}{\text{Velocity ratio}}.$$

If and, only if, a machine is perfectly efficient, M.A. = V.R.

$$\textbf{Percentage efficiency} = \frac{\textbf{M.A.}}{\textbf{V.R.}} \times \textbf{100.}$$

It is customary to calculate the V.R. of a machine in order that we may know the maximum M.A. of which it might be capable. In practice the value is often considerably below this value.

159. Pulleys

A brief reference has already been made to pulleys, but solely from the point of view of changing a direction. A combination of pulleys may constitute a very useful machine, however, and in this section we shall consider a variety of systems.

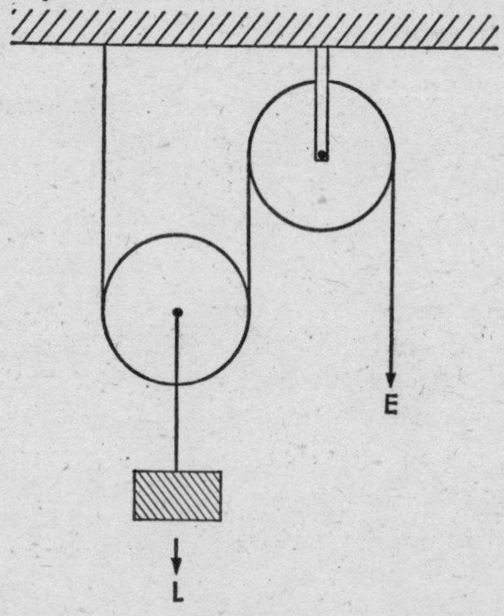

FIG. 144.

First consider a very simple type, which is illustrated in Fig. 144. One pulley is fixed, and merely serves to change the direction of the applied force. The other pulley is movable.

Calculation of velocity ratio.

Imagine that the load is raised s cm. There will therefore be $2s$ cm of slack rope.

$\therefore$ The effort pulls in $2s$ cm.

$$\text{V.R.} = \frac{2s}{s} = 2.$$

If it were possible to devise a perfect machine the M.A. would also be 2. Even without friction, however, the maximum of 2 could not be obtained, because the movable pulley would have some weight.

Suppose there were no friction, the tension in the rope would be the same throughout, and equal to E, the effort.

Forces on the movable pulley are:

$E + E$ upwards
$L \times W$ downwards, where $W =$ weight of pulley.

If the effort can just lift the load,

$$L = 2E - W$$

and
$$\text{M.A.} = \frac{2E - W}{E} = 2 - \frac{W}{E}.$$

We thus see that for M.A. to equal V.R. we have to have a weightless pulley as well as a frictionless system.

We also note that the actual M.A. varies with different values of E.

160. Pulley arrangements of the first order

Fig. 145 represents a pulley system of this type. There are 3 separate strings, 3 movable pulleys and 1 fixed pulley.

Calculation of V.R.

Imagine the load being raised a distance s cm. To do this $2s$ cm of slack must be taken up, i.e. pulley B must rise $2s$ cm.

If B rises $2s$ cm $2 \times 2s$ cm slack must be taken in from the relevant string, and this means that pulley C must rise $4s$ cm.

For this to happen the effort must draw in $8s$ cm.

$$\text{V.R.} = \frac{8s}{s} = 8 \text{ or } 2^3.$$

Had there been 4 movable pulleys the V.R. would have been 2^4, and for n pulleys the V.R. $= 2^n$.

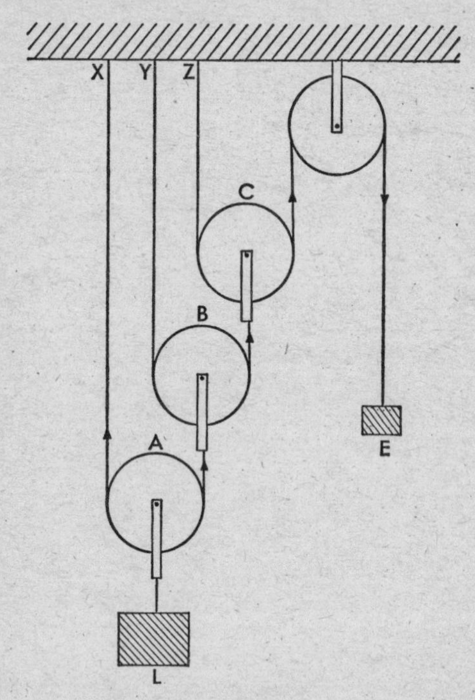

FIG. 145.

Once again this value of the V.R. presents only a highly theoretical maximum value of the M.A. which could be obtained in the absence of friction and with weightless pulleys.

The effect of friction cannot be calculated, but it is possible to calculate a maximum value for the M.A.

which allows for the weight of the pulleys. Let the weight of each pulley be W, and the weight of the load be L.

The forces on pulley A are

$L + W$ down

and $2\ T_a$ up, where T_a is the tension in the string XAB.

$$\therefore \qquad T_a = L/2 + W/2.$$

The forces on pulley B are

$T_a + W$ down

and $2\ T_b$ up, where T_b is the tension in YBC.

$$\therefore \qquad T_b = L/4 + 3W/4.$$

The forces on pulley C are

$T_b + W$ down

$2\ T_c$ up, where T_c is the tension in ZCE.

$$\therefore \qquad T_c = L/8 + 7W/8.$$

But $T_c =$ the effort,

$$\therefore \quad \text{M.A.} = \frac{L}{E} = \frac{L}{L/8 + 7W/8} = \frac{8}{1 + 7W/L}.$$

Thus, even in the absence of friction, the pulleys must be light compared with the load if the theoretical M.A. is to be approached.

This type of pulley system has a very high theoretical maximum for M.A. when there is a large number of pulleys, but, since most of the strings are of fixed length, it is difficult to get the load to rise very much unless the strings are very long, and this needs a lot of space.

161. Pulleys of the second system

This is the most commonly used system and it consists of two pulley "blocks". Each of these consists of a number of pulleys fixed together either as shown in Fig. 146(*a*) or, more usually, as in Fig. 146(*b*). The upper block is fixed, and the lower block is free to move.

FIG. 146(*a*). FIG. 146(*b*).

It makes no difference whether the individual pulleys within a block are co-axial or on top of each other. The whole arrangement is represented in Fig. 147.

Calculation of V.R.

Imagine the load L being lifted by s cm.

This is equivalent to removing 4 s cm of string.
Thus 4 s cm of string must be removed by the effort.

Thus $$\text{V.R.} = \frac{4s}{s} = 4.$$

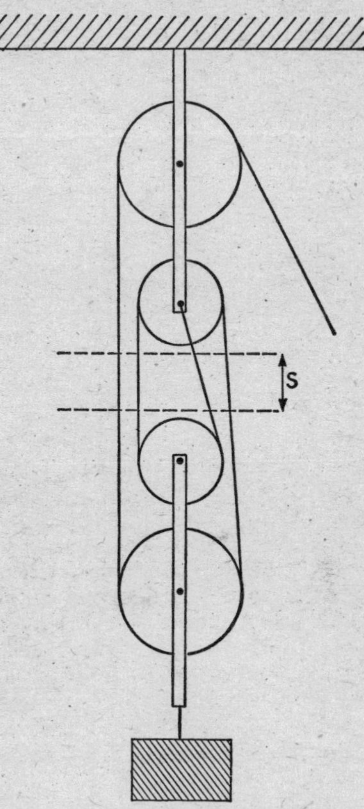

FIG. 147.

Had there been n strings the V.R. would have been "n".

Once again this represents a theoretical maximum value for the M.A. in the absence of friction and with weightless pulleys.

If the total weight of the lower pulley block was W,

and there was no friction, so that the same tension, E, equal to the effort, was felt in every part of the string, the forces on the lower block would be

$$4\,E \text{ up}$$
$$L + W \text{ down}$$

$$\therefore \qquad \text{M.A.} = \frac{4\,E + W}{E} = 4 + \frac{W}{E}.$$

162. The wheel and axle

This useful machine is shown in Fig. 148.

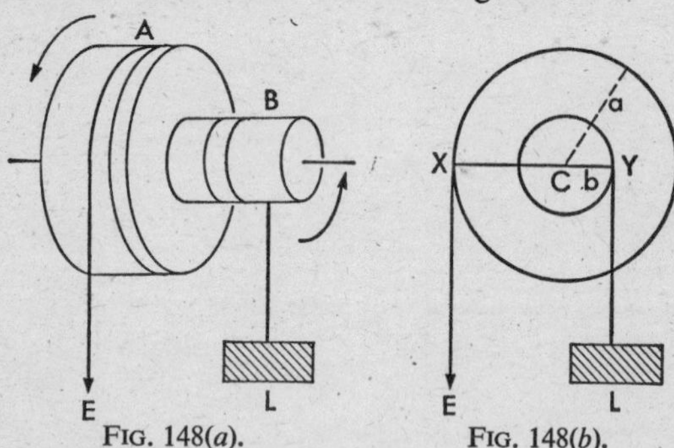

FIG. 148(*a*). FIG. 148(*b*).

It consists of:

(1) A large wheel (A), round which a rope is passed. The effort is then applied to the end of the rope. An alternative version applies the effort by means of a handle fixed at the perimeter of the wheel.

(2) The axle (B), cylindrical in shape and with its diameter smaller than that of the large wheel. A rope is fixed to it and wound round it in the **opposite** direction to that of the large wheel. The load is attached to the free end.

Calculation of V.R.

For each rotation of the large wheel there is a complete rotation of the axle. Thus, while the effort moves a distance equal to the circumference of the large wheel, the load is moved a distance equal to the circumference of the axle.

Let a be the radius of the large wheel and b be the radius of the small wheel.

Then, for one revolution,

$$\text{V.R.} = \frac{\text{Distance moved by effort}}{\text{Distance moved by load}} = \frac{2\pi a}{2\pi b} = \frac{a}{b}.$$

Again, if friction were absent, we would expect M.A. to equal V.R.

Fig. 148(b) represents an end view of the system, and XY can be regarded as a lever.

Taking moments about C, the centre,

$$E \times a = L \times b$$

or

$$\text{M.A.} = \frac{L}{E} = \frac{a}{b}.$$

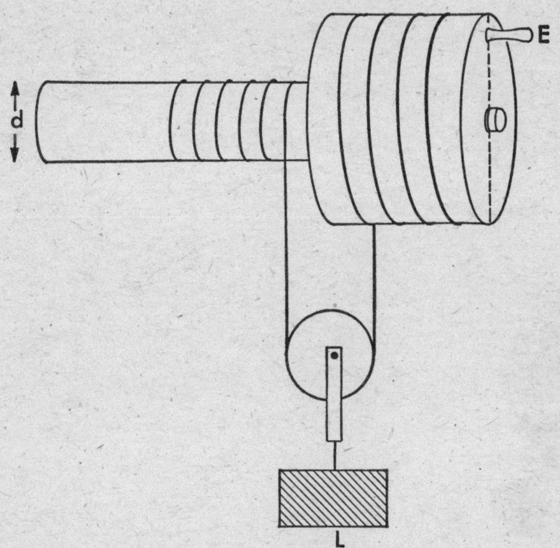

FIG. 149.

163. The differential wheel and axle

This is a variation of the wheel and axle with a high V.R. It is illustrated in Fig. 149 and it will be seen that the axle is in two parts, both of which are co-axial cylinders with different diameters D and d.

Calculation of V.R.

In one revolution E, the effort, moves round the circumference of the larger cylinder, and so moves a distance πD. Thus the length of rope taken up is πD. At the same time the smaller cylinder releases a length of rope $= \pi d$.

Thus the amount of rope removed is $\pi D - \pi d$

and the load is raised $\dfrac{\pi D - \pi d}{2}$.

Thus $$\text{V.R.} = \frac{\pi D}{\pi(D - d)} = \frac{D}{D - d},$$

and this can be extremely big as D is only slightly larger than d.

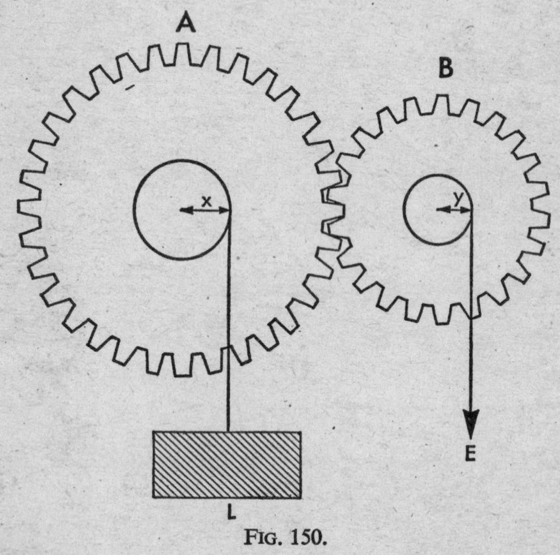

FIG. 150.

164. Gears

Fig. 150 represents two gear wheels with teeth that mesh. This means that they are the same size and the same distance apart round the rim of the wheel. For gear wheel A let there be n_A teeth round the perimeter, and let the radius of its axle be x.

For gear wheel B let there be n_B teeth round the perimeter, and let the radius of the axle be y.

We will assume the wheel B is turned by means of a rope wrapped round the axle, and that the force applied to the rope is E, the effort. Wheel A has a rope round its axle in the same direction, and it is this rope which lifts the load, L.

For one revolution of B, the effort must move $2\pi y$, and this causes n_B teeth to move. When wheel A, which has n_A teeth round its perimeter, is moved round by n_B teeth, it turns through $n_B \div n_A$ revolutions.

In one revolution the load is moved $2\pi x$,

$\therefore$ for n_B/n_A revolutions it moves $n_B/n_A \times 2\pi x$.

$$\text{Thus V.R.} = \frac{\text{Distance moved by effort}}{\text{Distance moved by load}} = \frac{2\pi y\, n_A}{n_B\, 2\pi x}$$
$$= \frac{n_A\, y}{n_B\, x}.$$

It is not possible to calculate the amount of friction there may be, but this gives an indication of the sort of M.A. that would be possible.

165. The inclined plane and the wedge

An inclined plane and its special form, the wedge, may be regarded as machines, since it is possible to use them in raising a large load by the application of only a small effort. In the inclined plane ABC (Fig. 151) let $AC = 1$ and let $BC = h$.

When a load is moved along the length of the plane, l, it is raised vertically through a height, h.

$$\therefore \qquad \text{V.R.} = \frac{l}{h}.$$

In this example there is likely to be a great deal of friction, so the M.A. will be much less than l/h. Nevertheless it is an extremely useful device which was used in the construction of the Pyramids, and which we use when we take a zig-zag path to climb a hill.

A wedge is a double inclined plane.

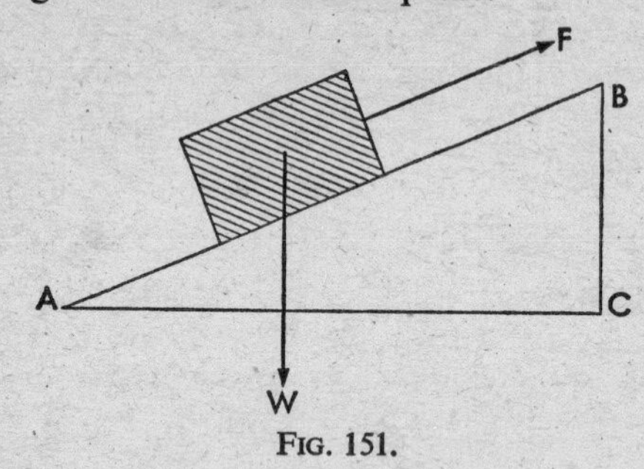

FIG. 151.

166. The screw

The screw makes use of the principle of the inclined plane. This can be seen by means of the following simple experiment.

Cut out in paper a right-angled triangle *ABC* (Fig. 152(*a*)). This represents the side view of an inclined plane.

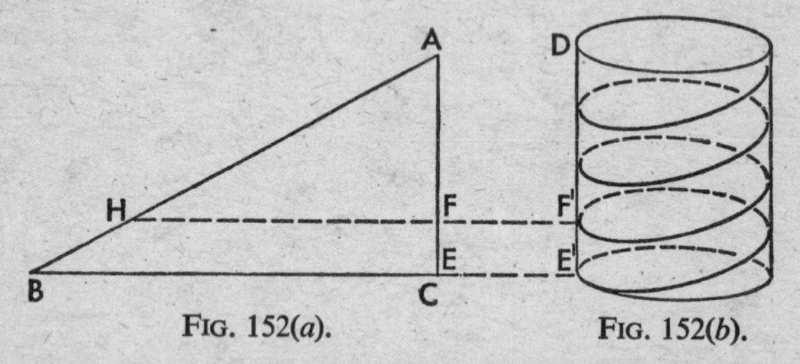

FIG. 152(*a*). FIG. 152(*b*).

Wrap this round a suitable cylinder, such as a pencil or a small cylindrical bottle. It will be seen that the edge *AB* will form a spiral curve round the cylinder in the same way that the thread of a screw winds round the screw shaft. This is represented in Fig. 152(*b*). The distance between two points, such as E' and F', where the spiral cuts a straight line DE', parallel to the axis of the cylinder, is called the "pitch" of the screw. The screw moves the load a distance equal to the pitch of the screw each time it is turned through one revolution, and it is this which enables the V.R. to be found.

In most cases the load is the force pulling two bodies, e.g. metal plates, together, but in some cases there is a rather more conventional lifting of a load. The frictional effect is usually very high.

167. The screw-jack

This machine uses the principle of the screw to lift heavy weights, e.g. a car. The effort is usually applied by means of a long-handled bar, which increases the V.R.

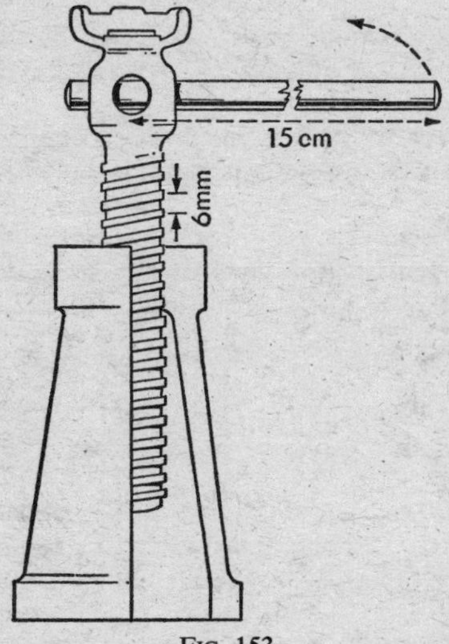

FIG. 153.

Consider a screw-jack with a pitch of 6 mm, which is turned by applying an effort to a bar 15 cm long (*see* Fig. 153).

Calculation of V.R.

When the bar turns through one revolution it moves through $2\pi \times 15$ cm.

The load is raised by 0·6 cm.

$$\therefore \quad \text{V.R.} = \frac{2\pi \times 15}{0{\cdot}6} = 50\pi \text{ or approx. } 150.$$

Even though the friction may be high there will be a high M.A., and this makes it very easy to lift a car when changing a wheel.

168. Experimental law for the machine

We have already seen in § 158–160 that, even in the absence of friction, the M.A. may vary for different efforts or loads.

In many cases it is important that we should know some equation which connects the effort with the load. Such an equation can, quite clearly, represent an empirical law only. It is not possible to produce an equation theoretically for we have no means of assessing the effect of friction. Nevertheless it is possible to use an experiment to obtain a series of effort and load values, which are then plotted on a graph. If the graph is a straight line, we say that the relationship between effort and load is a linear one, and the general form of the equation is

$$E = aL + b,$$

where E = Effort needed to lift a load L; "a" and "b" are constants.

If the value of the constants can be found by studying the graph, then the empirical law for the machine can be written down.

If, for example, we find that $a = 0{\cdot}56$ and $b = 1{\cdot}4$, then the law of the machine becomes

$$E = 0{\cdot}56\,L + 1{\cdot}4.$$

This can be used to find E when L is known, or vice versa. The constant "b" is the value of E when $L = 0$; it is the value of the resistance that has to be overcome to start the machine when there is no load.

An example of the application of the graphical method of obtaining a solution is given below. Other examples may be found in Vol. I of *National Certificate Mathematics* (published by the English Universities Press) and the student who has no experience of graphical methods is advised to study Chapter VIII of this book.

169. Worked example

In a series of experiments carried out with a Weston Differential Pulley, the effort E kgf necessary to raise a load of L kgf was found to be as follows.

L . . .	10	20	30	40	50
E . . .	3·3	4·8	6·4	7·9	9·5

Show these values on a graph and determine the law which they seem to follow, and find the probable effort when the load is 25 kgf.

Examining the data, and noting the maximum number to be plotted in each case, we can take 1 unit on the horizontal axis to represent 1 kgf for the load L, and 5 units to represent 1 kgf for the effort E.

Then plot the points as shown (Fig. 154). Since the data are derived from experimental results, slight deviations from a straight line are to be expected. If any one or two points are definitely not in accordance with the majority, they should be neglected.

A straight line should be drawn to take in as many of the points as possible, or, failing that, it should be so drawn that the points are fairly evenly distributed on either side of it.

We now take two points A and B on this line, which are suitable for reading off the values. They will not necessarily be either of the points actually plotted, and it is advisable to choose them fairly wide apart.

The quantities E and L are evidently connected by a linear law which will be of the form

$$E = mL + b.$$

For the point A, $L = 35$ kgf and $E = 7{\cdot}2$ kgf.
For the point B, $L = 12$ kgf and $E = 3{\cdot}6$ kgf.

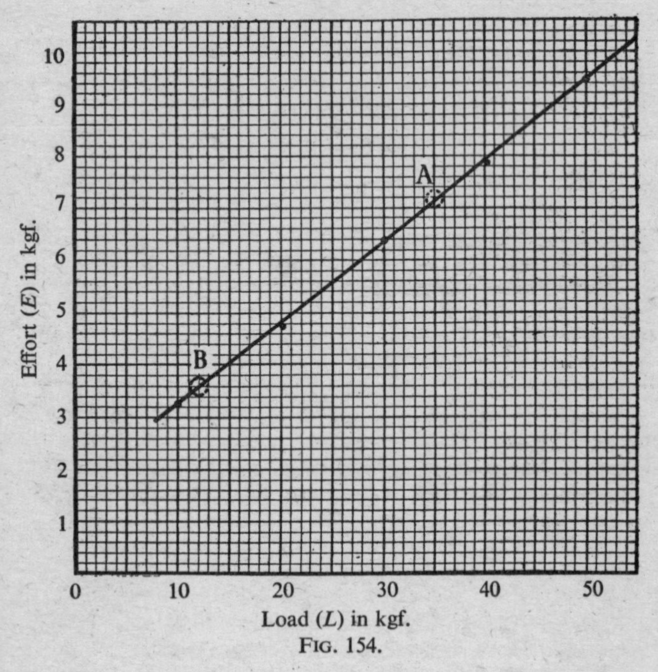

FIG. 154.

Hence, substituting in $E = mL + b$, because these values satisfy the required law, we have:

(1) $7{\cdot}2 = 35m + b$. (2) $3{\cdot}6 = 12m + b$.

Subtracting,

$$3{\cdot}6 = 23m$$

i.e.

$$m = \frac{3{\cdot}6}{23} = \frac{18}{115} = 0{\cdot}157$$
$$= 0{\cdot}16 \text{ approx.}$$

Substituting in (2),

$$3{\cdot}6 = 12m + b$$
$$3{\cdot}6 = \frac{12 \times 18}{115} + b.$$

$$\therefore \quad b = 3{\cdot}6 - 1{\cdot}88$$
$$= 1{\cdot}72.$$

Hence the law is $E = 0{\cdot}16L + 1{\cdot}72$.

To find E when the load is 25 kgf, substitute in this law thus determined.

Then $\quad E = 0{\cdot}157 \times 25 + 1{\cdot}72$
$\quad\quad = 5{\cdot}6$ kgf approx.

This result agrees very closely with the graph itself.

170. Algebraical method

The algebraical working employed in the graphical solution, described above, is the usual method of finding the law when two pairs of values of the load and effort are given.

If the law of the machine is known to be of the form $E = aL + b$, and two corresponding values of E and L are known, then on substituting these values in $E = aL + b$ two equations involving a and b as unknowns are obtained and these can be solved simultaneously.

171. Worked examples

Example 1. *The quantities E and L are connected by the equation* $E = aL + b$. *It is known that when*

$$L = 5, \quad E = 1{\cdot}5$$

and when $\quad L = 10, E = 3{\cdot}6.$

Find the law of the machine.

Substituting the corresponding values of E and L in

$$E = aL + b$$

we get $\quad 1{\cdot}5 = 5a + b \quad . \; . \; . \; . \; . \; . \quad (1)$

and $\quad 3{\cdot}6 = 10a + b \quad . \; . \; . \; . \; . \; . \quad (2)$

Subtracting $\quad 2{\cdot}1 = 5a.$

$\therefore \quad a = 0{\cdot}42.$

Substituting this value for (a) in equation (2), we get:

$$3{\cdot}6 = 10 \times 0{\cdot}42 + b,$$

whence $\quad b = 3{\cdot}6 - 4{\cdot}2.$

$\therefore \quad b = -0{\cdot}6.$

Substituting these values for a and b in $E = aL + b$, we get:

$$\mathbf{E = 0{\cdot}42L - 0{\cdot}6.}$$

This is the law of the machine.

Example 2. *The velocity ratio of a screw-jack is* 60. *Its efficiency is* 30 *per cent. What effort is needed to lift a load of* 1 *tonne force with this jack?*

$$\begin{aligned} 1 \text{ tf} &= 10^3 \text{ kgf} \\ &= 9{\cdot}8 \times 10^3 \text{ N}. \end{aligned}$$

Let E Newtons be the effort required.

$$\text{Percentage efficiency} = \frac{\text{M.A.}}{\text{V.R.}} \times 100$$

$$\therefore \qquad 30 = \frac{\text{M.A.}}{60} \times 100$$

whence $\qquad \text{M.A.} = 18.$

But $\qquad \text{M.A.} = \frac{L}{E}$

$$= \frac{9800}{E}$$

$$\therefore \qquad 18E = 9800$$

$$\therefore \qquad E = 545 \text{ N}.$$

172. Summary

A summary of important formulæ is printed below for convenient reference when working problems in the exercise which follows:

(1) Velocity Ratio

$$= \frac{\textbf{distance moved by applied force}}{\textbf{distance moved by load}}.$$

(2) Mechanical Advantage

$$= \frac{\textbf{load}}{\textbf{applied force}}.$$

(3) Per cent. Efficiency

$$= \frac{\textbf{mechanical advantage}}{\textbf{velocity ratio}} \times \mathbf{100.}$$

Exercise 23

1. In a machine, a mass of 80 kg was moved through 3 m by an effort of 250 N, which moved through 10 m. Find: (*a*) the velocity ratio (V.R.); (*b*) the mechanical advantage (M.A.); (*c*) the percentage efficiency.

2. In a machine, a load of 160 N is lifted by an effort of 28 N; the V.R. is 8. What is the percentage efficiency?

3. In a machine, the load is 400 N and the efficiency at that load is 64 per cent. If the V.R. is 5, what is the applied force?

4. It is necessary to lift a load of 600 N by a screw-jack in which the efficiency, at that load, is 25 per cent. If the V.R. is 150, find the effort and the M.A.

5. A tonne is drawn through 0·25 km up an inclined plane of slope 1 in 30. The effort was 500 N parallel to the plane. What is the efficiency of the plane, regarded as a machine?

6. A screw-jack has 90 threads to the metre; the effort is applied at the end of an arm 30 cm long. What force must be applied to lift a load of 120 kg, if the efficiency at that load is 25 per cent?

7. In a screw-jack the effort, E, required to lift a load, L, is applied at the end of a handle which describes a circle of 20 cm radius. The pitch of the screw is 1 cm. What is the V.R.? What is the efficiency if an effort of 5 N will lift a load of 112 N $\left(\pi = \frac{22}{7}\right)$?

8. In an experiment on an inclined plane it was found that a force of 65 N, applied up the plane and parallel to it, was necessary to haul a 10-kg mass slowly up the plane. If the inclination of the plane is 30° and its length is 6 m, find (*a*) the work done against gravity; (*b*) the work done against friction when the mass is pulled the whole length of the plane; (*c*) the efficiency of this plane, considered as a simple machine.

In experiments carried out with various machines, the effort and the load were measured; they are set out in the tables below. For each machine it may be assumed that the law connecting E and L is of the form $E = aL + b$, where a and b are constant. E and L are always measured in the same units. Plot the values on a graph, and deduce the most accurate forms of the law.

9.

L . . .	30	40	60	70	80
E . . .	2·13	2·6	3·8	4·3	5·1

10.

L . . .	14	42	84	112
E . . .	5·1	13·3	26·0	35·3

11.

L .	7	14	21	28	35	42	49	56
E .	3	6·5	9·5	12·5	16	18·75	22	25

12.

L . . .	100	120	140	160	180	200
E . . .	12·6	13·8	15·7	17·6	19·6	21·5

CHAPTER XV

COMPOSITION OF VELOCITIES; RELATIVE VELOCITY

173. Composition of velocities

Just as a body may be acted upon by more than one force, so it may have imparted to it more than one velocity. A simple example may make this clear.

Suppose a man to be moving across the deck of a moving ship at right angles to the axis of the ship, and therefore at right angles to the direction in which the ship is moving.

Let the man start from O (Fig. 155) and at the end of 2 s have moved to A_1, where $OA_1 = a$ m.

During the 2 seconds the ship will have moved a distance relative to the sea in the direction OB. Let this distance be represented by OB_1, where $OB_1 = b$ m.

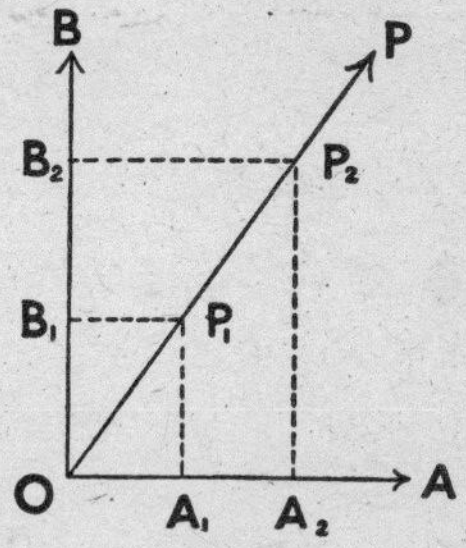

FIG. 155.

These two displacements OA_1 and OB_1 in the directions OA and OB respectively are equivalent to a single displacement OP_1 in the direction OP. Relative to the sea the man will have moved through OP_1.

Similarly in the next 2 seconds. Suppose the man to have moved to A_2 in the direction OA and to B_2 in the direction OB. The man's position relative to the sea will be P_2 and he will thus have moved through P_1P_2, assuming both his velocity and that of the ship to be uniform.

Thus in 4 seconds the man will have moved, relative to the sea, through the distance OP_2. Therefore, he has a velocity along OP which will be equal to $\frac{OP_2}{4}$ m per second.

Using the same term which is adopted when a body is acted upon by two forces, we can say that the velocity of the man along *OP* is the **resultant of the velocities** along *OA* and *OB*.

The velocity of an aeroplane is a striking example of the composition of velocities. It possesses the velocity due to the engines in the direction set by the aviator and, in addition, a velocity which is imparted by the force of the wind. This must be ascertained in magnitude and direction before it is possible to learn the exact direction and velocity with which the aeroplane is moving.

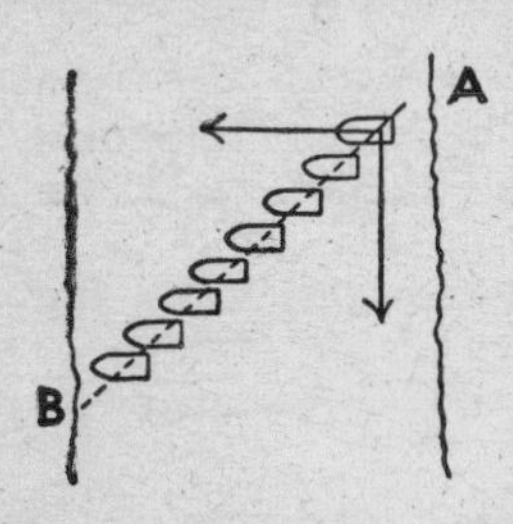

FIG. 156.

There is another instance in which the actual direction of the resultant velocity is apparent to an observer. Suppose a man starts to row a boat across a stream with a strong current and points his boat directly at the opposite bank (*see* Fig. 156). The current imparts to his boat an additional velocity down-stream and at right angles to the velocity which he gives to the boat as a result of his rowing. As a consequence of having these two simultaneous velocities, he actually moves with a resultant velocity, as shown in Fig. 156, the direction of which will be along a straight line such as *AB*. It would be a matter of calculation, which will be discussed in this chapter, in what direction he must point the boat if he is to arrive at a point on the other bank directly opposite to *A*.

174. The Parallelogram of Velocities

The student will be prepared to learn that, just as when two forces act on a body their resultant can be found by means of the law of the Parallelogram of Forces, so if a body receives two velocities, these have a **resultant velocity** which can be determined by a corresponding parallelogram law. This is known as the Parallelogram of Velocities, and is defined as follows:

Parallelogram of Velocities

When a body has two simultaneous velocities and these are represented in magnitude and direction by two adjacent sides of a parallelogram, their resultant velocity is represented in magnitude and direction by the diagonal of the parallelogram which passes through their point of intersection.

The Parallelogram of Forces was established as a result of experiments, and no mathematical proof was given. Experimental demonstration of the Parallelogram of Velocities, however, is not easy or satisfactory, and it is not proposed at this stage to burden the student with a mathematical proof. The parallelogram law applies to all vector quantities, and its truth will be assumed.

175. Calculation of a resultant velocity

The resultant of two velocities is found in the same way as the resultant of two forces.

Let *OA*, *OB* (Fig. 157) represent two velocities imparted to a body.

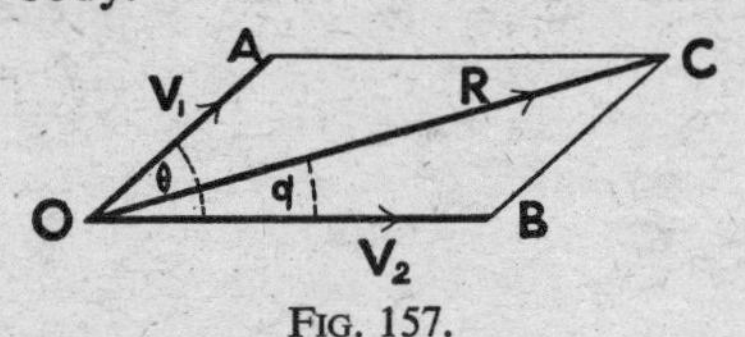

FIG. 157.

Let θ be the angle between them.
Complete the parallelogram *OACB*.
Draw the diagonal *OC*.
Then *OC* represents the resultant of the velocities represented by *OA* and *OB*.

(1) **To find the value of** *OC*. Using the rule proved in § 49

$$OC^2 = OA^2 + OB^2 + 2OA \,.\, OB \cos AOB.$$

Let v_1 and v_2 represent the velocities.
Let V be their resultant.

Then, on substitution

$$V^2 = v_1^{\,2} + v_2^{\,2} + 2v_1v_2 \cos \theta.$$

(2) **To find α,** the angle between the resultant and v_2, as in § 49:

$$\tan \alpha = \frac{v_1 \sin \theta}{v_2 + v_1 \cos \theta}.$$

Note. It is a good plan to check the working by finding also the angle between v_1 and the resultant. If this be β, then:

$$\tan \beta = \frac{v_2 \sin \theta}{v_1 + v_2 \cos \theta}.$$

When the directions of the velocities are at right angles. As will be seen from Fig. 158,

$$\text{Resultant velocity} = \sqrt{v_1^{\,2} + v_2^{\,2}}.$$

Also

$$v_2 = V \cos \theta$$
$$v_1 = V \sin \theta.$$

176. Component velocities

A velocity can be resolved into two component velocities by the method employed to obtain components of a force. The most useful case is that in which we require to obtain two components at right angles. The method again is the same as for forces, and will be evident from an examination of Fig. 158.

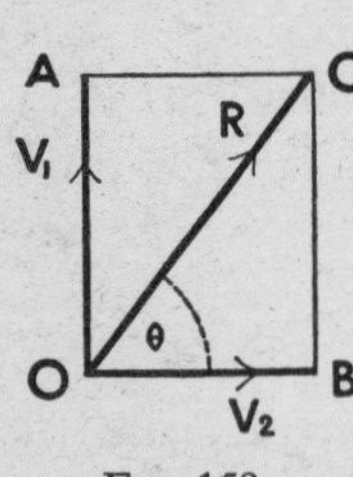

FIG. 158.

If V be the velocity of which it is desired to find components making angles of θ and 90° − θ with it, the components will be:

$$V \cos \theta, \quad V \sin \theta.$$

177. Worked example

In order that the pilot of an aeroplane may reach his destination at a specified time, it is necessary, after making allowances for the strength and direction of the wind, that he should travel at 360 *km/h at an angle* 30° *E. of N. The wind is blowing from the south at* 80 *km/h. In what direction and with what velocity must he set his course?*

Note. As was the case when dealing with problems on forces, problems may usually be solved either (1) by accurate drawing, or (2) by trigonometry.

In this problem it will be seen that we know the magnitude and direction of the **Resultant velocity**—viz. 360 km/h at 30° E. of N. We also know **one of the components**—viz. the velocity of the wind, 80 km/h from the south.

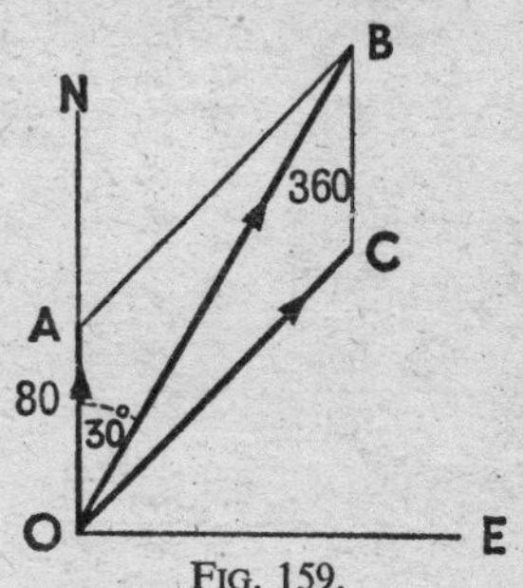

FIG. 159.

Construction

Draw OA on a suitable scale to represent velocity of wind (Fig. 159).

Draw OB making 30° with OA and to represent 360 km/h on the scale chosen.

This represents the Resultant velocity.

Join AB.

Complete the parallelogram $OABC$.

Then OC **represents the other component**—i.e. the magnitude of the velocity with which the aviator starts and the direction in which he points the machine.

We need therefore to find:

(1) The magnitude of OC.

(2) The angle it makes with the north or east.

This can be done by means of a carefully drawn figure or

We can obtain the solution by trigonometry, as follows:

It is necessary to solve $\triangle OBC$ or $\triangle OAB$.

This is the case of "two sides and an included angle".

The cosine rule may be used, or if we wish to employ logs, the formula

$$\tan\frac{B-C}{2} = \frac{b-c}{b+c}\cot\frac{A}{2}$$

is better.

The solution by this means is given as an example. Using the form

$$\tan \frac{A - B}{2} = \frac{a - b}{a + b} \cot \frac{C}{2},$$

let $\angle OAB = A$,
then $a = 360$.
Let $\angle ABO = B$,
then $b = 80$.
Let $\angle AOB = C$.

$\therefore$ Substituting in the above formula

$$\tan \frac{A - B}{2} = \frac{360 - 80}{360 + 80} \cot \frac{30°}{2}$$
$$= \frac{280}{440} \cot 15°.$$

Taking logs

$$\log \tan \frac{A - B}{2} = \log 140 + \log \cot 15° - \log 220$$
$$= 0{\cdot}3756$$
$$= \log \tan 67°\ 10'.$$

No.	log
280	2·4471
cot 15	0·5719
	3·0190
440	2·6434
log tan 67° 10′	0·3756

$\therefore \quad \frac{A - B}{2} = 67°\ 10'$

and $\quad A - B = 134°\ 20'$.
Also $\quad A + B = 150°$.
Adding $\quad 2A = 284°\ 20'$.
$\therefore \quad$ **$A = 142°\ 10'$.**
Subtracting $\quad 2B = 15°\ 40'$.
$\therefore \quad$ **$B = 7°\ 50'$.**
$\therefore \quad \angle AOC = 30° + 7°\ 50'$
$\quad = 37°\ 50'$.

$\therefore$ **Direction of course is 37° 50′ E. of N.**

To find *AB*:

Using the sine rule:

$$\frac{AB}{\sin 30} = \frac{80}{\sin 7°\ 50'}.$$

$$\therefore \quad AB = \frac{80 \sin 30°}{\sin 7° 50'} = \frac{40}{\sin 7° 50'}$$

$[\sin 30° = \frac{1}{2}]$.

Using logs as shown in working

$AB = 293{\cdot}6$ or 294 approx.

No.	log
40	1·6021
sin 7° 50′	$\bar{1}$·1345
293·6	2·4676

∴ The pilot must set his course at 37° 50′ E. of N. at 294 km/h approx.

Exercise 24

1. Find the resultant of each of the following pairs of velocities, giving its direction relative to the larger component:

(*a*) 12 m/s and 8 m/s; angle between them 60°;
(*b*) 10 km/h and 12 km/h; angle between them 40°.

2. A ship is steaming at 15 km/h and a man walks across its deck at 3 km/h, in a direction at right angles to the axis of the ship. What is the man's resultant velocity relative to the sea?

3. A ship is heading west at 12 knots. A current deflects it so that it actually travels in a direction 20° N. of W. at 15 knots. What is the velocity of the current?

4. A river 2 km wide flows with a velocity of 1 km/h. A man swimming with a speed of 2 km/h crosses the river at right angles to the bank. How long does he take?

5. The pilot of an aeroplane which flies at 200 km/h, when there is no wind, wishes to travel due north when a west wind (i.e. a wind from the west) is blowing at 40 km/h. Find, graphically, the direction in which he must point the nose of the machine and his speed relative to the ground.

6. The time allowed for an aeroplane to fly to a particular town is 3 h. The town is 480 km due east of the starting point and the wind is blowing from the south-east at 60 km/h. What velocity must the pilot give the plane?

7. An aeroplane has to fly 600 km due north. Its normal speed is 300 km/h and a wind is blowing from the

south-west at 60 km/h. How long must the pilot allow for the trip, flying at his normal speed?

8. Two velocities at right angles to each other are in the ratio 3:4 and the magnitude of their resultant is 60 km/h. What are the two speeds?

178. Relative velocity

It will be helpful in understanding what is termed relative velocity if the student will recall some of the sensations which he has probably experienced when, travelling in a train, another train has been moving on the next set of rails.

(1) Suppose that the trains were moving in the **same direction,** yours at 40 km/h, the other at 50 km/h. The other train appears to be moving very slowly; sometimes, even if the running is very smooth, the other train may seem to stand still while yours is moving backwards. The motion which the other train seems to possess is called the **relative velocity** of the other train with respect to yours. The velocity with which the other train seems to be moving will clearly depend on the difference between the velocities of the two trains. If the velocities were as stated above, and it were possible for you to measure the velocity with which the other train seems to move—i.e. its relative velocity—this would be 10 km/h in the direction in which you are both going.

To a man in the other train, your train, on the contrary, will appear to be moving backwards at 10 km/h. This would be the relative velocity of your train with respect to the other.

It should be noted that **if the velocities of both trains were to be increased by the same amount,** the relative velocities would be the same. If, for example, each were to be increased by 20 km/h to 60 km/h and 70 km/h respectively, the relative velocity would still be 10 km/h.

(2) Suppose the trains were moving in **opposite directions,** yours moving north at 40 km/h, the other moving south at 50 km/h. At the end of an hour, if the trains continued to move uniformly, your train would have travelled 40 km north, while the other would have gone

50 km south. The two trains would be (40 + 50), i.e. 90, km apart. They would have separated at the rate of 90 km/h.

Each train is moving relatively to the other at 90 km/h. This is therefore the **relative velocity.**

Now let us approach the question in another way. Suppose that **each train were to receive an additional velocity equal and opposite to that of the north-bound train**—viz. 40 km/h.

Then the train going north would have no velocity—i.e. it would be at rest.

But the velocity of the south-bound train would be equal to 40 + 50, or 90 km/h south.

Then, to the north-bound train the south-bound train would appear to be moving south at 90 km/h.

∴ **The relative velocity of the south-bound train with reference to the one going north is 90 km/h.**

In a similar way the relative velocity of the north-bound train with reference to the train going south could be found.

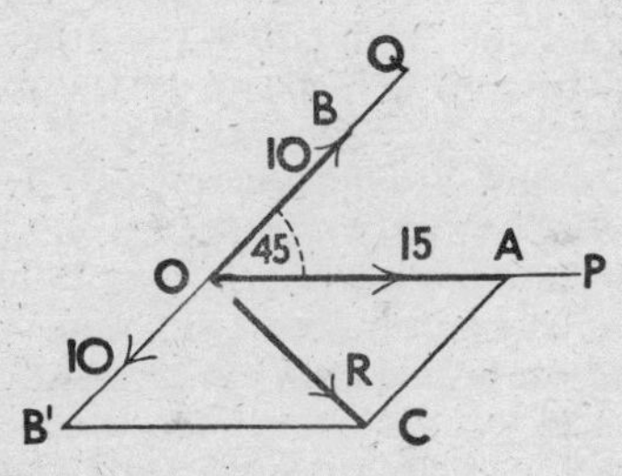

FIG. 160.

179. Relative velocities of bodies starting from the same point and diverging

Suppose that two cyclists P and Q start from a point O (Fig. 160) and travel as follows:

P along OP at 15 km/h,
Q along OQ at 10 km/h.
How does P appear to Q to be moving?

Or **what is the relative velocity of *P* with respect to *Q*?**

Let us adopt the device which was employed in the case of trains moving along parallel lines.

The method may be stated thus:

Apply to each a velocity equal and opposite to the velocity of one of them, say Q. Then Q will be brought to rest. The resultant velocity of P will be the relative velocity of P with respect to Q.

In accordance with this we apply to each a velocity equal and opposite to *Q*'s velocity—which is 10 km/h along *OQ*.

Let *OB* represent *Q*'s velocity.

Let *OA* represent *P*'s velocity.

Then *OB'* equal and opposite to *OB* represents the velocity added to each.

As a consequence *Q* will be at rest.

The velocity of *P* will be the resultant of

15 km/h along *OP*—i.e. *OA*

and 10 km/h along *OB'*.

Complete the parallelogram *OACB'*.

Then the diagonal *OC* represents in magnitude and direction the resultant of the two velocities represented by *OA* and *OB'*.

Or ***OC* represents in magnitude and direction the relative velocity of *P* with respect to *Q*.**

This may be measured or calculated as shown in § 173.

As an exercise the student should draw the construction for obtaining the **relative velocity of *Q* with respect to *P*.**

This will be done by applying to each a velocity equal and opposite to the velocity of *P*.

Definition. *The relative velocity of P with respect to Q is the velocity, both in magnitude and direction, with which P appears to Q to be moving.*

180. Relative velocity of bodies not diverging from the same point

The following example will illustrate this case and indicate the method of solution.

A ship, A (see *Fig.* 161), *is moving due north at* 15 *knots.*

A ship, B, 10 *miles due north, is moving west at* 10 *knots. What is the relative velocity of* A *with respect to* B*?*

The positions of the ships are indicated in Fig. 161.

To the velocity of each add a velocity equal and opposite to that of *B*—i.e. we add a velocity of 10 knots east.

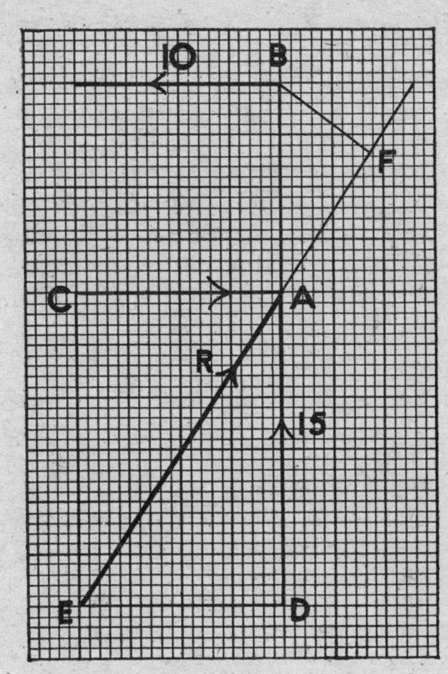

FIG. 161.

Then *B* is relatively at rest.

The resultant of the two velocities now imparted to *A* is the resultant of the velocities represented by *DA* and *CA*.

Complete the rectangle *ACED*. Then the diagonal *EA* represents the resultant of the two velocities.

$\therefore$ ***EA* represents in magnitude and direction the relative velocity of *A* with respect to *B*.**

These can now be determined by drawing or by calculation.

By calculation

$$\begin{aligned} AE &= \sqrt{15^2 + 10^2} \\ &= \sqrt{325} \\ &= \mathbf{18} \text{ nearly.} \end{aligned}$$

Also $$\tan CAE = \frac{15}{10} = 1{\cdot}5.$$

$\therefore$ $$\angle CAE = 56° \, 19'.$$

$\therefore$ The relative velocity of A with respect to B is 18 knots in a direction 56° 19′ N. of E.

It should be noted that if EA be produced, since it represents the path of A relative to B, by drawing BF perpendicular to EA, **the length of BF represents the shortest distance between A and B as they move on their respective paths.**

Exercise 25

1. Two men, A and B, meet at crossroads and proceed up two different roads, A going due north and B going due east, at 3 km/h and 4 km/h respectively. What is the velocity of B relative to A?

2. A ship, A, is sailing E. 30° N. at 12 km/h. A second ship, B, is moving S. 30° N. at 9 km/h. What is the velocity of A relative to B?

3. A train A travels due north at 30 km/h and a train B due west at 40 km/h. What is the velocity of A relative to B?

4. A car goes due east at 30 km/h and from it a train appears to be moving due north at 60 km/h. What is the true velocity of the train?

5. A hovercraft is 10 km due south of a yacht. The hovercraft is moving north at 12 km/h and the yacht is sailing west at 9 km/h. What is the velocity of the hovercraft relative to the yacht? Find, from a scale drawing, how near the hovercraft approaches the yacht.

6. A body is moving on bearing 0° with speed 20 m/s

and its velocity relative to another body appears to be 30 m/s on bearing 45°. Find the velocity of the second body.

7. A ship sails west with a speed of 12 knots while another steams south-east at a rate of 18 knots. Find, graphically, the velocity of the second ship relative to the first.

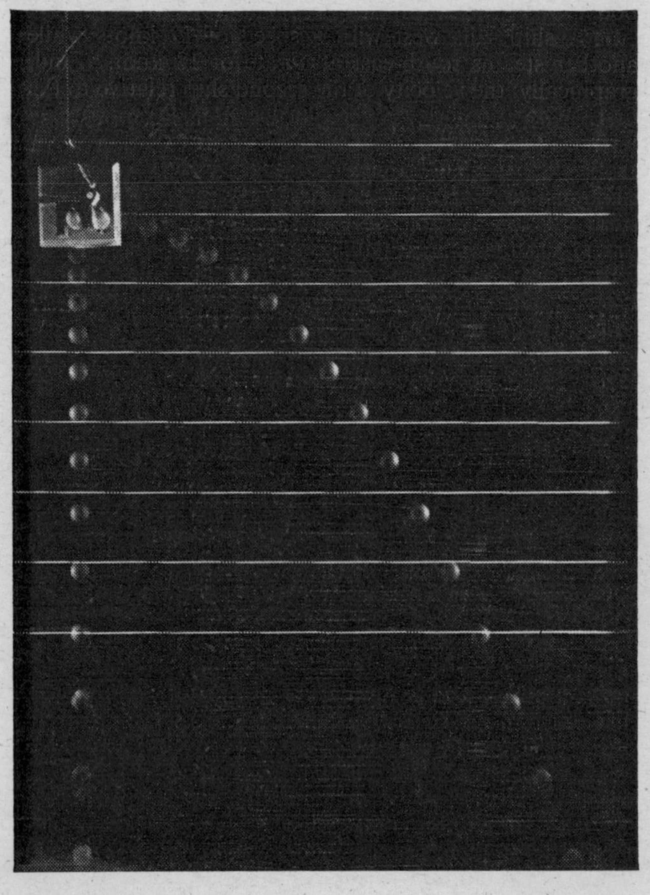

FIG. 162. Path of projectile displayed by stroboscopic photography. (From *PSSC Physics*, D. C. Heath and Company, Lexington, Mass., 1965.)

CHAPTER XVI

PROJECTILES

181. The path of a projectile

Everybody is familiar with the fact that when a cricket ball, or a rocket, or a jet of water from a hose-pipe is impelled into the air, the path which it describes is a curve.

A photograph of such a trajectory, taken by multi-flash photography, is shown in Fig. 162.

This curve is a **parabola,** and the path described is called the **trajectory.**

In each case there is an impelling force whose direction is inclined to the vertical. If it were otherwise we know that a body projected vertically upwards moves in a straight line to a highest point and then descends vertically to its starting point.

Let V denote the velocity imparted to a projectile by an impelling force.

From the moment that the body begins to move upwards it is acted upon by the force of gravity and its velocity decreases.

Consider its position after t s.

Under the action of the impelling force, it will have moved a distance Vt in the direction in which it was projected (represented by OA in Fig. 163).

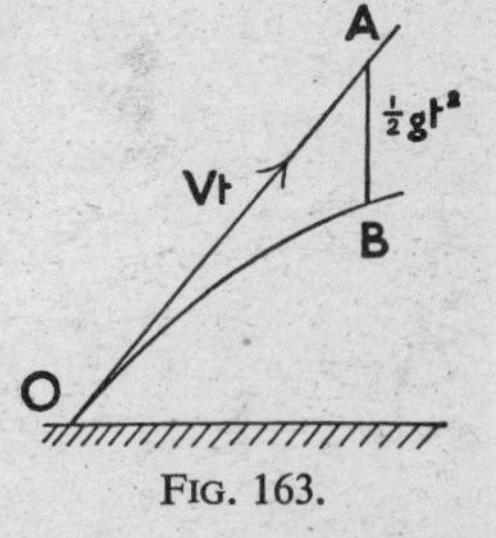

FIG. 163.

Under the action of gravity, during the interval, it will have fallen a distance $\frac{1}{2}gt^2$ vertically downwards, represented by AB.

If a series of values of t were taken, corresponding points, such as B, would be obtained, all of which lie on the curve.

182. Components of the initial velocity

Let V be the velocity of projection.

Let α be the angle of projection made with the horizontal.

Resolve V into vertical and horizontal components.

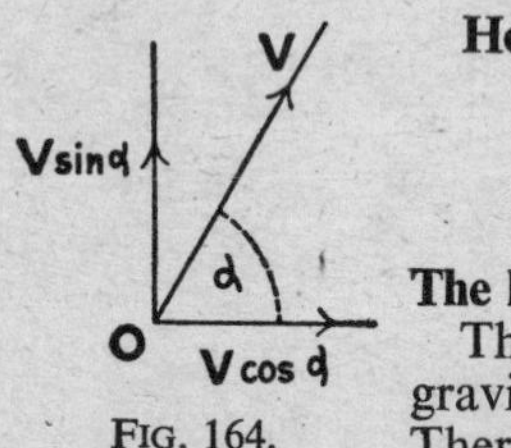

FIG. 164.

Horizontal component = $V \cos \alpha$ (Fig. 164).

Vertical component = $V \sin \alpha$.

The horizontal component, $V \cos \alpha$.

This is unaffected by the force of gravity, being at right angles to it. Therefore:

It remains constant throughout the flight.

∴ After any time t, **distance covered = $V \cos \alpha \times t$.**

The vertical component, $V \sin \alpha$.

This is subject to a retardation due to the force of gravity.

∴ After time t, it becomes $V \sin \alpha - gt$.

The **vertical distance** described after time t is

$$V \sin \alpha \times t - \tfrac{1}{2}gt^2.$$

(1) When $gt = V \sin \alpha$, i.e. the vertical component is equal to zero, the body projected is at the **highest point of the curve** (A in Fig. 165).

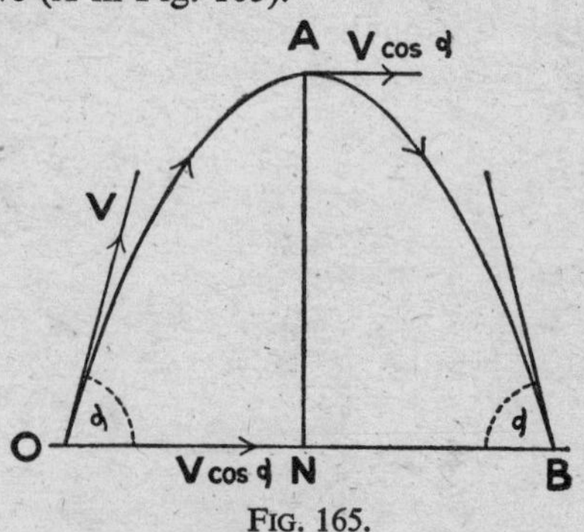

FIG. 165.

At that point the only velocity which the body has is the horizontal component, $V \cos \alpha$.

(2) After the highest point the body moves along the downward part, its vertical component subject to the acceleration due to gravity and with horizontal velocity, $V \cos \alpha$ (Fig. 165).

(3) The body reaches the ground again at B with a velocity which is the resultant of its horizontal and vertical velocities,

viz. $$V \cos \alpha \text{ horizontal},$$
$$-V \sin \alpha \text{ vertical}.$$

Consequently the velocity with which it strikes the ground is the same in magnitude as the initial velocity—viz. V.

The angle which V makes with the horizontal is α, measured this time in the opposite direction (Fig. 165).

The curve is thus a symmetrical one, and the vertical line AN through the highest point is the axis of symmetry.

The above results are not exact in practice, since they are affected by air resistance.

183. Formulæ connected with projectiles

(1) Greatest height

(*a*) To find the time

At the moment of the greatest height, the vertical component vanishes:

$$\therefore \quad V \sin \alpha + gt = 0 \qquad (see\ \S\ 192)$$

$$\therefore \quad t = \frac{V \sin \alpha}{g}.$$

(*b*) To find the greatest height

When a body is projected vertically upwards, h, the highest point, is given by:

$$v^2 = 2gh \quad . \quad . \quad . \quad (\S\ 103)$$

or $$h = \frac{v^2}{2g}.$$

$\therefore$ At the greatest height of the projectile

$$h = \frac{V^2 \sin^2 \alpha}{2g}.$$

As a special case, when

$$\alpha = 45°.$$

Since $\sin 45° = \sqrt{\tfrac{1}{2}}, \quad h = \dfrac{V^2}{4g}.$

(2) Time to describe the range

As shown above, time to reach the greatest height is given by

$$t = \frac{V \sin \alpha}{g}.$$

But time for the whole flight must be twice this.
$\therefore$ for the whole range

$$t = \frac{2V \sin \alpha}{g}.$$

(3) To find the range

The distance OB (Fig. 165) is called the range.
Let R represent the range.
Since the horizontal component is unchanged; in time t,

$$R = V \cos \alpha \times t.$$

But, for the complete range

$$t = \frac{2V \sin \alpha}{g} \qquad (\textit{see above}).$$

$\therefore$
$$R = V \cos \alpha \times \frac{2V \sin \alpha}{g}$$
$$= \frac{2V^2 \sin \alpha \cos \alpha}{g}$$

or
$$R = \frac{V^2 \sin 2\alpha}{g},$$

since $\sin 2\alpha = 2 \sin \alpha \cos \alpha.$

Maximum range for a given velocity of projection.
From the above

$$R = \frac{2V^2 \sin \alpha \cos \alpha}{g},$$

where R is the range.

Since $\sin \alpha = \cos (90° - \alpha)$
and $\cos \alpha = \sin (90° - \alpha)$.

∴ The same values of R will be obtained whether the angle of projection is α or $90° - \alpha$.

For example, the range will be the same whether the angle of projection is 30° or 90° − 30°—i.e. 60°.

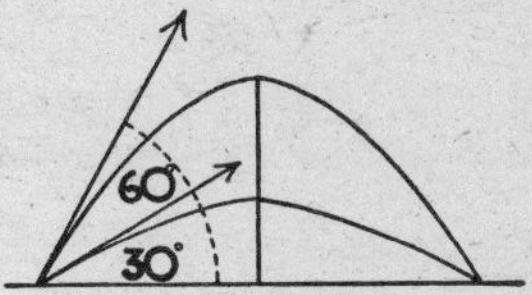

Although the range will be the same for both angles, the time taken and the greatest height will be different. This is apparent from Fig. 166.

FIG. 166.

For the larger angle, V **sin** α, the vertical component will be greater than V **cos** α, the horizontal component, since **sin** α increases when α increases. For the smaller value of α, V **cos** α will be the greater, since $\cos \alpha$ decreases as α increases.

Maximum range

Using the formula

$$R = \frac{V^2 \sin 2\alpha}{g}.$$

If V be constant, R is greatest when $\sin 2\alpha$ is greatest. But the maximum value of the sine of an angle is unity, when the angle is 90°.

∴ $\sin 2\alpha$ is greatest when $2\alpha = 90°$
and $\alpha = 45°$.

Then $$R = \frac{V^2}{g}.$$

When, therefore, in projecting a body, the maximum range is desired, as in throwing a cricket ball, the body should be projected at 45° to the horizontal.

184. Motion of a body projected horizontally from a height

A familiar phenomenon of modern times is that of a body being dropped from an aeroplane, whether it be a package of food for the succour of marooned men, or a bomb intended for a more sinister purpose.

Suppose that, as shown in Fig. 167, a body is dropped from an aeroplane at A, a distance of h m above the ground, it being assumed the plane is flying horizontally.

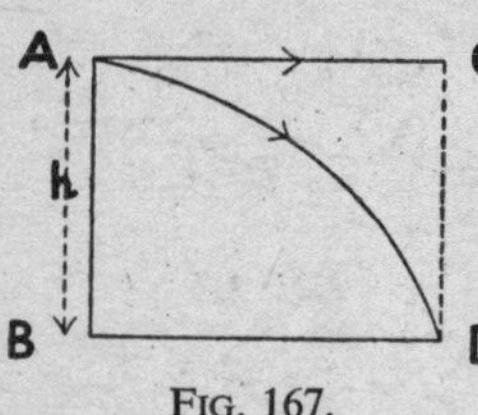

Fig. 167.

Let V = velocity of the plane.

At the moment of projection the body projected will have the same velocity as the aeroplane —i.e. V in a horizontal direction. Therefore in their further motions the aeroplane will continue to move in a horizontal direction with velocity V, while the velocity of the bomb will have a horizontal component V. The bomb will also have a downward vertical velocity due to gravity, with an acceleration g.

$\therefore$ Time taken to reach the ground will be the same as if it fell from a height h.

The time of fall is given by the formula $h = \frac{1}{2}gt^2$.

$\therefore$ In time t given by this equation the bomb will have:

(1) Fallen a distance represented by AB or CD.

(2) Travelled a horizontal distance represented by AC, where

$$AC = V \times t.$$

$\therefore$ Its final position will be given by D, and the path described is a parabola.

Since the aeroplane also moves with a horizontal velocity V, in the same time t it will have travelled a distance equal to $V \times t$ and represented by AC. Thus it will have reached the point C, vertically above D, the position of the bomb.

This assumes that there is no air resistance affecting the motion of either body, but in practice, although this is not very considerable at low velocities, it is far from negligible at high velocities.

Both will also be affected by wind and air currents.

185. Summary of formulæ

1. **Components of velocity of projection**

Horizontal	$V \cos \alpha$.
Vertical	$V \sin \alpha$.

2. **Height,** after time "t"

$$h = Vt \sin \alpha - \tfrac{1}{2}gt^2.$$

3. **Time** of flight $\quad t = \dfrac{2V \sin \alpha}{g}.$

4. **Horizontal range** $\quad R = \dfrac{V^2 \sin 2\alpha}{g}.$

5. **Maximum range** $\quad R = \dfrac{V^2}{g}$
   when $\quad \alpha = 45°.$

6. **Greatest height** $\quad h = \dfrac{V^2 \sin^2 \alpha}{2g}.$

186. Worked example

The pilot of an aeroplane flying at a height of 4900 *m, with a velocity of* 360 *km/h, aims at bombing a target directly in front of him. At what distance away, reckoned horizontally, must he release the bomb, if air resistance and air currents be ignored?*

$$360 \text{ km/h} = 100 \text{ m/s}$$

Time for bomb to fall from 4900 m is found by using the formula $\quad s = \tfrac{1}{2}gt^2.$

Substituting $\quad 4900 = 4{\cdot}9t^2.$

$\therefore \quad t^2 = 1000$

and $\quad t = \mathbf{10\sqrt{10}}$ **s.**

Distance travelled horizontally in this time is

$$100 \times 10\sqrt{10} \text{ m}$$
$$= 1000\sqrt{10} \text{ m}$$
$$= \sqrt{10} \text{ km}$$
$$= 3{\cdot}16 \text{ km.}$$

∴ He must release the bomb at a distance of 3·16 km (approx.), measured horizontally from the target.

Exercise 26

1. A body is projected with velocity 80 m/s at 30° with the horizontal. What is its range and greatest height reached? ($g = 10$ m/s^2.)

2. Find the range of a gun when the muzzle speed is 400 m/s and the elevation 24·5°.

3. A target is 10 km from a gun and the gun's muzzle speed is 400 m/s. What angle of projection will ensure a hit?

4. A projectile has a muzzle speed of 70 m/s. What is the range on a horizontal plane if the angle of projection is 15°. Find the error, at this range, if the error in the elevation is 0·25°.

5. A projectile has a muzzle speed of 700 m/s. What is its maximum horizontal range?

6. A stone is thrown upwards with a speed of 35 m/s and at an elevation of 45°. When and where does it reach the ground?

7. A cricket ball is thrown at an elevation of 45° and pitches 20 m from the thrower; his height above the ground being negligible, and the ground being horizontal, what was the ball's initial speed, and how high does it rise in the air?

8. A shot is fired with a speed of 200 m/s at elevation 30°. How high is it after 5 s? What is the greatest height it reaches and what is the horizontal range?

9. A stone is thrown horizontally with speed 70 m/s from the edge of a cliff 40 m high. What is the time which elapses before the stone reaches the ground and how far is the horizontal distance of the point of projection from the point of impact?

10. A plane is flying at 1960 m at 630 km/h when it drops a bomb. Find (*a*) the time taken for the bomb to reach the ground; (*b*) the horizontal distance between the point of release and the point of impact.

11. A projectile is fired with velocity 300 m/s at 15° to the horizontal; it just hits its target, and g is 9·8 m/s^2. What would the error have been if g had been taken as 10 m/s^2? What percentage error is this?

CHAPTER XVII

DENSITY AND RELATIVE DENSITY

187. Density

An old trick question used to be to ask "Which weighs most, a kg of lead or a kg of feathers?" If the mass of both is the same, and they are weighed at the same place, their weights, i.e. the forces with which they are attracted by the earth, will be the same. Their volumes, however, will be vastly different. Specimens of lead, wood and polystyrene can all have the same mass, but their volumes will be very different.

When we talk about lead being "heavier than wood" we are implying that the same volume is considered in both cases. We are really commenting on the way their densities compare.

The density of a substance is a measure of the mass found in a given volume, e.g. in 1 cubic metre.

$$\textbf{Density} = \frac{\textbf{Mass}}{\textbf{Volume}}.$$

Hence:

$$\textbf{Mass} = \textbf{Volume} \times \textbf{Density}, \qquad \textbf{Volume} = \frac{\textbf{Mass}}{\textbf{Density}}.$$

The mass of 1 cubic metre of four common substances is shown below.

Wood (Pine)	Aluminium	Concrete	Iron
500 kg	2700 kg	1600 kg	7870 kg

Thus the density of aluminium is $2{\cdot}7 \times 10^3$ kg/m^3.

188. Density and the weight of an object. Example

If steel has a density of $7{\cdot}7 \times 10^3$ *kg/m*^{3}*, find the weight of a rolled steel joist* 10 *m long, which has a total cross-sectional area of* 100 *cm*^{2}.

$$1 \text{ cm} = 10^{-2} \text{ m.}$$
$$\therefore \quad 1 \text{ cm}^2 = 10^{-4} \text{ m}^2$$
$$\therefore \quad 100 \text{ cm}^2 = 10^{-2} \text{ m}^2.$$
$$\begin{aligned} \text{Volume} &= \text{cross-sectional area} \times \text{Length} \\ &= 10^{-2} \times 10 \\ &= 0{\cdot}1 \text{ m}^3. \\ \text{Mass} &= \text{Volume} \times \text{Density} \\ &= 0{\cdot}1 \times 7{\cdot}7 \times 10^3 \\ &= 770 \text{ kg.} \\ \text{Weight} &= M\,g \\ &= 770 \times 9{\cdot}8 \\ &= 7550 \text{ N.} \end{aligned}$$

The weight might also be expressed as 770 kgf.

189. Relative density

It is sometimes useful to make use of the relative density.

$$\text{Relative density } (d) = \frac{\text{Density of substance}}{\text{Density of water}}$$

$$\text{alternatively } d = \frac{\text{Mass of a given volume of a substance}}{\text{Mass of same volume of water}}$$

or,

since weight $= m \times g$,

$$d = \frac{\text{Weight of given volume of substance.}}{\text{Weight of same volume of water}}$$

It will be noticed that relative density is a ratio. Thus it has no limits. An alternative (older) name is "specific gravity".

190. Worked example

The relative density of a type of concrete is 1·5. *Find the mass of concrete that would be needed to fill a trench* 40 *cm wide,* 25 *cm deep, and* 4 *m long. The density of water is* 1 t/m^3.

Working in metres, the volume of the trench is

$$= 0{\cdot}4 \times 0{\cdot}25 \times 4$$
$$= 0{\cdot}4 \text{ m}^3.$$

Density = Relative density × Density of water

$\therefore$ for concrete, density $= 1{\cdot}5 \times 1 \text{ t/m}^3$.

$$\text{Mass} = \text{Volume} \times \text{Density}$$
$$= 0{\cdot}4 \times 1{\cdot}5$$
$$= 0{\cdot}6 \text{ t}$$
$$= 600 \text{ kg}.$$

Exercise 27

(*The density of water is* 10^3 kg/m^3.)

1. A block of lead has volume 314 cm^3 and mass 4·3 kg. What is its density in kg/m^3? What is its relative density?
2. A block of ice is 2 cm × 3 cm × 3·5 cm and has relative density 0·92. What is its mass?
3. The relative density of milk is 1·03. A bottle of milk was analysed and found to have relative density 1·027. What percentage of the liquid is added water?
4. One l of a liquid of $d = 0{\cdot}7$ is mixed with 1·5 l of a liquid of d = 1·3. What is the relative density of the mixture, assuming that there is no loss of volume involved in the mixing?

5. A container full of water weighs 5·01 N. The empty container has 10·1 N of sand added to it, and is then completely filled up with water. The new weight is 12·01 N. What is the relative density of the sand?

CHAPTER XVIII

LIQUID PRESSURE

191. Pressure defined

Even a moderately heavy bag becomes painful to hold by its handle if that handle is narrow. It is not the total weight that is responsible for the pain, but the fact that the weight is distributed over a small area. A relatively small force (or weight) concentrated on a small area may have a very big effect. Thus a stiletto heel can cause severe damage not only to a dance floor but also to the floor of an aircraft. The quantity which measures the extent to which a force is spread out over an area is called pressure.

$$\textbf{Pressure} = \frac{\textbf{Force}}{\textbf{Area over which it is applied}}$$

$$\text{or } \mathbf{P} = \frac{F}{A}.$$

It is usually measured in Newtons per square metre (N/m^2), but it could be measured by any other unit of force divided by any other unit of area. Thus "tonnes-force per sq kilometre" might be used as a unit of pressure, but it would be rather unusual.

A stiletto heel might be only 1 cm^2 in area, and at one moment might be supporting a weight of 50 kgf which is about 490 N.

$$1\ cm^2 = 0{\cdot}0001\ m^2$$

$$\therefore \text{ Pressure} = \frac{\text{Force}}{\text{Area}} = \frac{490}{0{\cdot}0001} = 4\,900\,000\ N/m^2$$

$$= 4{\cdot}9\ MN/m^2.$$

In contrast a sandal might have an area of 100 cm^2, so that the pressure would be 49 000 N/m^2.

The pressure from the stiletto heel is 100 times that

from a sandal. Small wonder that the airline companies were concerned!

Sometimes, of course, the effect can be used to advantage. A thin piece of wire can be used as a knife to cut cheese. A knife itself concentrates the force on to the very small area, which is the area of the cutting edge. The actual edge is often less than 0·001 mm thick, so that a knife, 4 cm of which is in contact with the material to be cut, presents an area of only 0·04 mm², so the pressure from even a relatively small force is very large.

Alternatively we may use a large area to spread a force more widely. This is the principle behind the use of snowshoes and of the use of large tyres on vehicles which have to move over soft ground.

192. Worked example

A section of earth can withstand a pressure of 1·5 *MN/m² without yielding. Calculate the minimum cross-sectional area of the plate to be fastened at the bottom of a metal post if the post has to carry a load of* 600 *kg.*

$$1{\cdot}5\ \text{MN/m}^2 = 1{\cdot}5 \times 10^6\ \text{N/m}^2.$$

The force on the post is

$$600 \times 9{\cdot}8 = 5900\ \text{N (approx.)}.$$

If the area of cross-section is A m²

$$\text{Pressure} = \frac{5900}{A},$$

and this must be $< 1{\cdot}5\ \text{MN/m}^2$.

$$\therefore \quad \frac{5900}{A} < 1{\cdot}5$$

or
$$\frac{5900}{1{\cdot}5 \times 10^6} < A,$$

i.e.
$$A > 3930 \times 10^{-6}\ \text{m}^2$$

or
$$A > 39{\cdot}3\ \text{cm}^2.$$

Exercise 28

(*The density of water is* 1000 *kg/m*^{3}.)

1. What pressure does a man exert if he is of mass 70 kg and he is standing on skis of area 0·3 m^2 each?

2. A 3-tonne tractor rests on four wheels, each of which has an area of 125 cm^2 in contact with the ground. What pressure does it exert?

3. A force acts over an area of 0·4 m^2, creating a pressure of 30 N/m^2. What is the force?

4. A man can produce a force of 12 N and wishes to create a pressure of 3 kN/m^2. What area must the force act over?

5. What is the pressure at the base of a column of water which is of circular cross-section, radius 2 mm and 220 mm high?

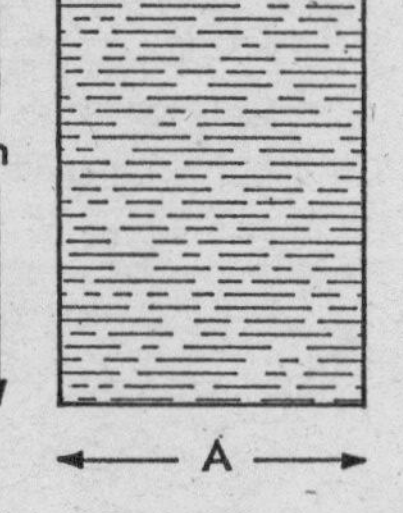

FIG. 169.

193. Pressure from a column of liquid

Fig. 169 represents a column of liquid with a uniform cross-sectional area, A sq m. Let the height of the column be h metres and the density of the liquid be ρ kg/m^3.

The volume of the liquid is $h \times A$ m^3.

$\therefore$ The mass of the liquid is $h \times A \times \rho$ kg.

$\therefore$ The weight of the liquid is $hA\rho$ kgf or $hA\rho g$ Newtons.

This weight is distributed over A sq m.

$$\therefore \text{Pressure} = \frac{F}{A} = \frac{hA\rho}{A} \text{ kgf/m}^2 \qquad \text{or} = \frac{hA\rho g}{A} \text{ N/m}^2.$$

i.e. Pressure caused by a liquid column h metres high

$$= h\rho \text{ kgf/m}^2 \quad \text{or} \quad h\rho g \text{ N/m}^2.$$

This is, of course, only the pressure provided by the liquid column. If there were also some pressure on the top surface of the liquid in Fig. 169 the pressure at the bottom of the column would be greater.

194. The direction of pressure

Pressure at a given point in a liquid acts equally in all directions.

A simple demonstration to illustrate this statement is shown in Fig. 170 which represents a simple syringe,

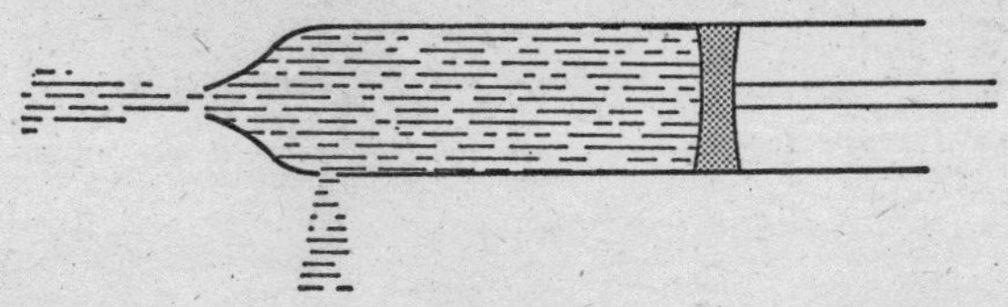

FIG. 170.

viewed from the top, which has a hole in the side as well as at the end. It is found that when the piston is pushed in the water squirts just as far to the side as it does forwards. This is a consequence of the mobility of the tiny particles which make up the liquid.

195. Movement within liquids

Consider two columns of the same liquid, connected by a thin tube, as shown in Fig. 171. Let there be a small

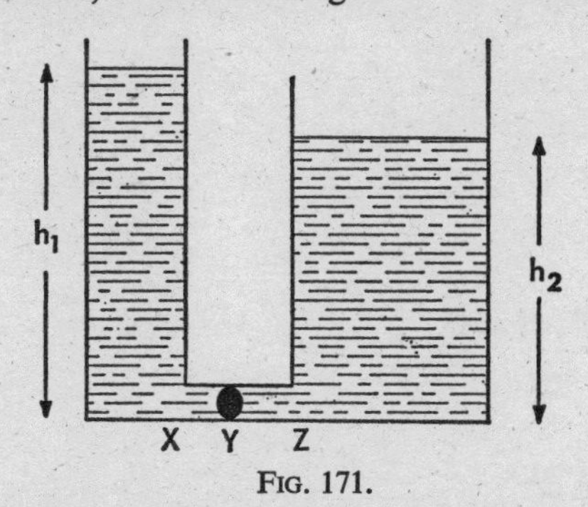

FIG. 171.

index Y in this tube, and let it be free to move. In practice this might be a pellet of mercury.

The pressure at $X = h_1 \rho g$.

Pressure acts in all directions equally, ∴ this is the pressure on the left-hand side of Y.

Let the cross-sectional area of the connecting tube be A.

∴ Force on left-hand side of Y is $h_1 \rho g \times A$.
Similarly the pressure at $Z = h_2 \rho g$.

∴ Pressure on right-hand side of $Y = h_2 \rho g$.

∴ Force on right-hand side of $Y = h_2 \rho g A$.

Thus Y feels a resultant force $= h_1 \rho g A - h_2 \rho g A$.
Thus there will be an acceleration of Y unless or until $h_2 = h_1$.

This is often summed up by saying

Liquids find their own level.

Alternatively we may say:

The pressures at all points on the same level in a liquid at rest are equal.

196. Pressure in a conical flask

At a point X (Fig. 172) we would expect the pressure to be $h_1 \rho g$.

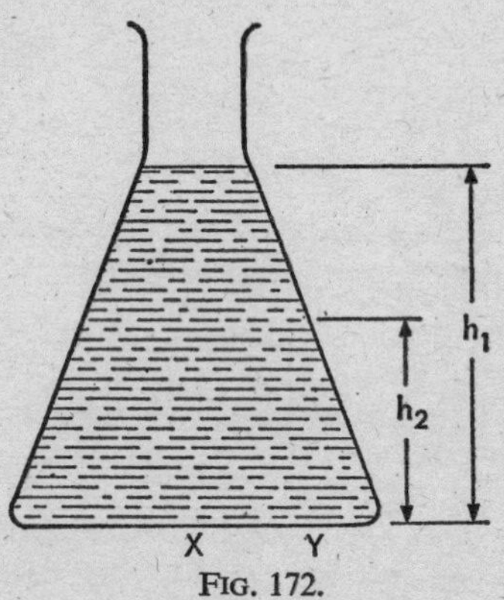

FIG. 172.

On the other hand, at Y there is not such a large column of liquid above it. We might be tempted to assume that the pressure was $h_2 \rho g$. If this were the case, however, the pressures at X and Y would be different, and the greater pressure at X would compress the liquid at Y until the pressures were the same.

The pressure caused by a column of liquid is

$$h\rho g$$

where h is the vertical height;

ρ is the density;

g is the relationship between mass and the force from gravity.

This pressure is independent of the actual shape of the container. In Fig. 173 the pressures at X, Y, and Z are all equal.

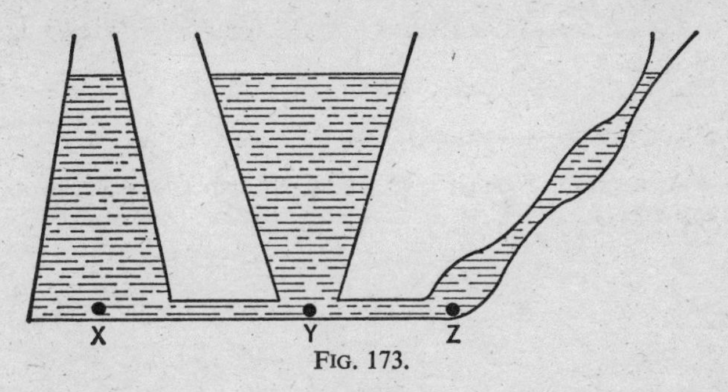

FIG. 173.

197. Average thrust from liquid pressure. Dams

The pressure in a liquid increases with depth, and it can become very large indeed. As the pressure increases the thrust on 1 square metre increases, and it is for this reason that dams are constructed with bases much thicker than their tops.

Consider a dam 40 m deep and 100 m wide. The pressure at the bottom provided by the water

$$= h\rho g$$
$$= 40 \times 1000 \times 9{\cdot}8 \quad \text{(density of water} = 1000 \text{ kg/m}^3\text{)}$$
$$= 392\,000 \text{ N/m}^2.$$

Since the pressure increases uniformly with depth it is fair to consider the average pressure in working out the total thrust.

Average pressure $= \frac{1}{2} \times 392\,000$ N/m^2.

The area of the dam over which the pressure acts

$$\begin{aligned} &= 100 \times 40 \\ &= 4000 \text{ m}^2. \\ \text{Total thrust} &= \text{Pressure} \times \text{Area} \\ &= 196\,000 \times 4000 \\ &= 784\,000\,000 \text{ N} \\ &= 784 \text{ MN}. \end{aligned}$$

(Alternatively. 784 000 000 N = 80 000 000 kgf
= 80 000 tonnes force.)

198. Worked example

A hole of area 5 cm^2 in the bottom of a boat has been plugged by a cork. If the hole is 60 cm below the water line, calculate the force trying to push the cork out of the hole.

60 cm below the surface of the water the pressure from the water is

$$\begin{aligned} &= h\rho g \\ &= 0{\cdot}6 \times 1000 \times 9{\cdot}8 \\ &= 5900 \text{ N/m}^2 \\ &= 5{\cdot}9 \text{ kN/m}^2. \\ \text{Thrust} &= \text{Pressure} \times \text{Area} \\ 5 \text{ cm}^2 &= 5 \times 10^{-4} \text{ m}^2. \\ \therefore \quad \text{Thrust} &= 5{\cdot}9 \times 10^3 \times 5 \times 10^{-4} \\ &= 2{\cdot}8(5) \text{ N}. \end{aligned}$$

199. Transmission of pressure

In § 195 it was shown that if two parts of the same liquid at the same level were at different pressures, there would be movement in the liquid. If the liquid is not free to move the result is that the particles making up the liquid are pushed together *slightly* more tightly, so that the pressure increases. The word "slightly" is important, for liquids are almost incompressible. There is, however, an increase in pressure.

Fig. 174 represents an L-shaped tube, with a piston at the lower end. Initially the pressure at $X = P_1$ $(= F_1/A)$. Y is at the same level,

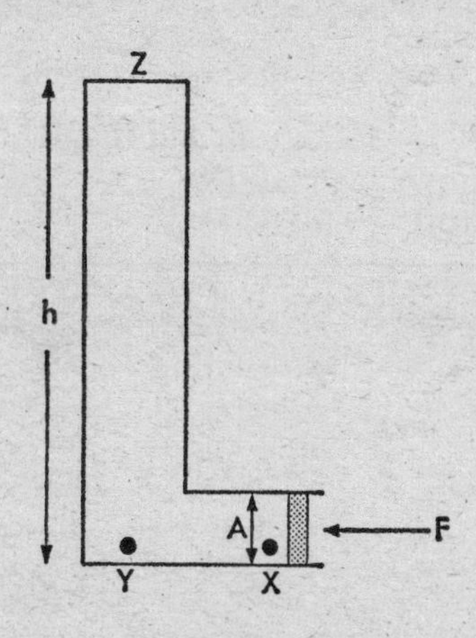

FIG. 174.

$\therefore$ Pressure at $Y = P_1$.

The pressure at Y is $>$ pressure at Z by $h\rho g$.

$\therefore$ Pressure at $Z = P_1 - h\rho g$.

Let the force at X be increased by F_2

$\therefore$ Pressure becomes $P_1 + P_2 = (F_1 + F_2)/A$.

$\therefore$ Pressure at Y becomes $P_1 + P_2$

and pressure at Z becomes $P_1 + P_2 - h\rho g$.

The same sort of argument can be applied to all parts of the liquid and it will be seen that the pressure at each place rises by P_2.

If a liquid is at rest any change in pressure at one point is transmitted equally to all other points within the liquid.

This principle is the basis on which the hydraulic braking system of a car works. It is illustrated in Fig. 175. Pushing down the brake pedal causes an increase in the pressure of the brake fluid, and this results in a greater thrust pushing the brake shoes against the inside of the brake drum.

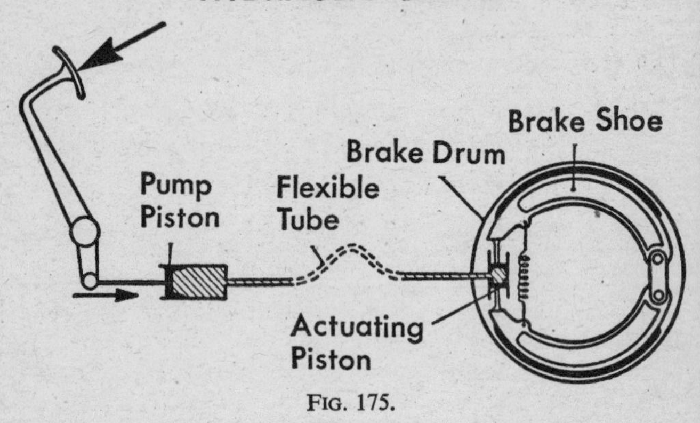

FIG. 175.

200. The hydraulic lift

The ability of liquids to transmit pressure is also used in the hydraulic lift, which is illustrated in Fig. 176.

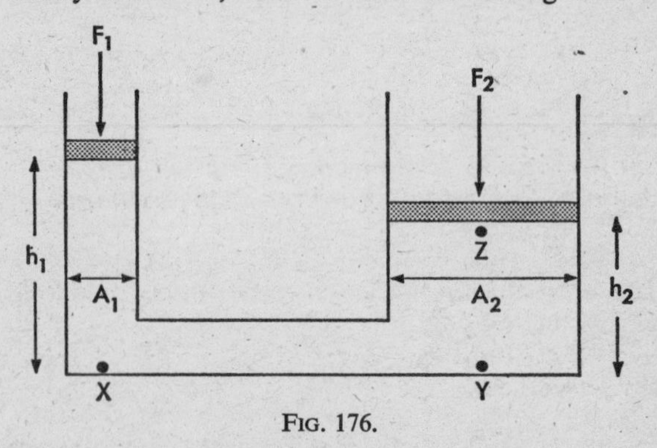

FIG. 176.

Suppose a force F_1 is applied to a piston of area A_1 resting on top of the liquid in the narrow limb.

$\therefore$ Pressure exerted on the top of the liquid

$$= \frac{F_1}{A_1}.$$

The pressure at $X = \frac{F_1}{A_1} + h_1\rho g.$

This is also the pressure at Y.

The pressure at $Z =$ pressure at $Y - h_2\rho g$.
$\therefore$ if the force at Z is acting on an area A_2 the force at

$$Z = \text{pressure at } Z \times A_2 = F_2.$$

$$\frac{F_2}{A_2} = \text{pressure at } Z = \frac{F_1}{A_1} + h_1\rho g - h_2\rho g.$$

In most cases $h_1\rho g - h_2\rho g$ is insignificant in comparison with F_1/A, so that

$$\frac{F_1}{A_1} = \frac{F_2}{A_2}.$$

This device is a machine which can be used for converting a small force, F_1, into a larger force, F_2. In the absence of considerations about friction we have shown that

$$\text{Mechanical advantage} = \frac{L}{E} = \frac{F_2}{F_1} = \frac{A_2}{A_1}.$$

Calculation of V.R.

Let the load rise a distance s.
$\therefore$ Volume of liquid pushed into right-hand limb

$$= s \times A_2.$$

This must have been pushed out of the left-hand limb by the effort.

If the distance moved by the effort is S, the volume of liquid pushed out

$$= S \times A_1.$$

$$\therefore \quad S \times A_1 = s \times A_2.$$

$$\text{V.R.} = \frac{S}{s} = \frac{A_2}{A_1}.$$

The disadvantage of this sort of hydraulic lift is that it would need a very long column of liquid on the effort side if there was to be a reasonably high V.R. (and hence, we hope, a high M.A.).

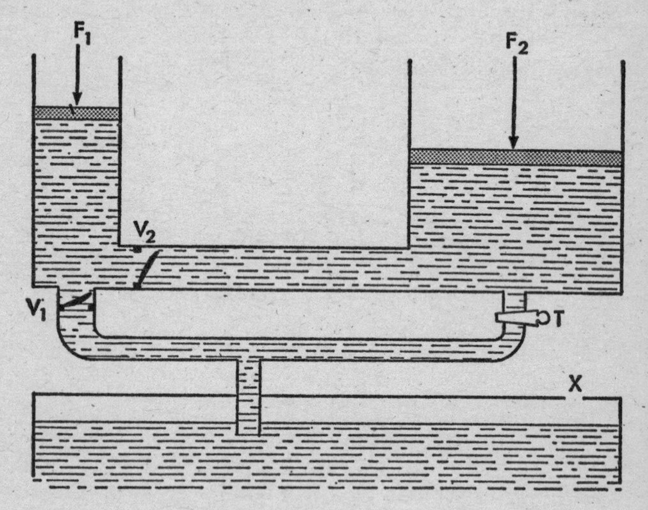

FIG. 177.

201. A practical hydraulic lift

Fig. 177 shows how the difficulty mentioned above can be overcome by use of a reservoir of liquid, two valves V_1 and V_2, and a return tap T. A description of its mode of action is deferred until § 215 in Chapter XIX.

202. Archimedes' Principle

Heavy objects are much easier to lift when they are in water than when they are in air. It is worth testing the validity of this statement by tying a piece of string round, say, a house brick, and then immersing it in water. As the brick is lifted out of the water by means of the string the tension in the string suddenly increases.

If we analyse the forces acting on the brick when it is in the water we find that there is

(i) A downward gravitational force—mg.
(ii) An (upward) force from the tension in the string—T.

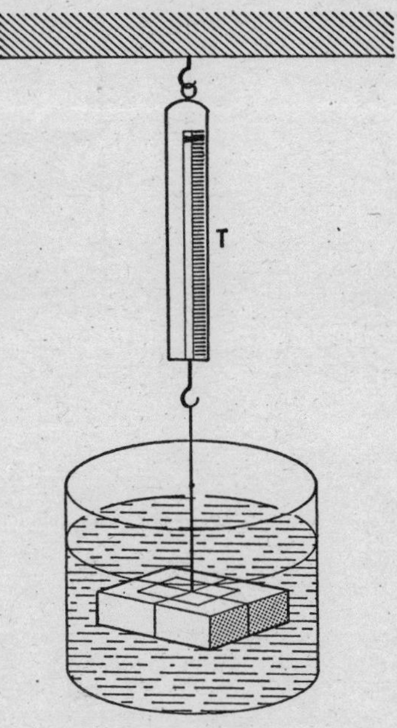

Fig. 178.

Experimentally we find that $T < mg$, and yet the object is not accelerating downwards. There must, therefore, be a third force, the upthrust, which, presumably, comes from the liquid itself.

If the size of this upthrust is measured in a series of simple experiments it is found that:

When a body is wholly or partially immersed in a fluid it experiences an upthrust equal to the weight of the fluid displaced.

This is known as **Archimedes' Principle.**

The word fluid is used because the "Principle" applies equally well to both liquids and gases.

203. Archimedes' Principle and pressure

Consider a body with uniform cross-sectional area A m², of which h metres is submerged in a liquid of density ρ kg/m³ (Fig. 179).

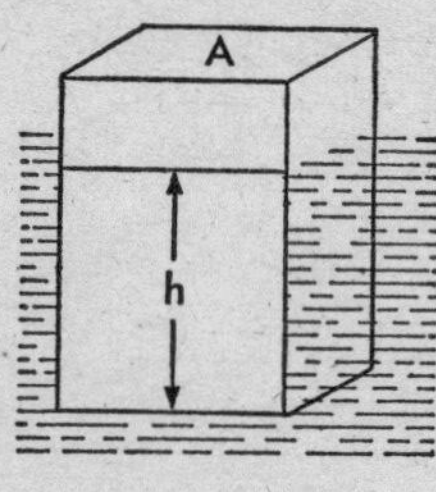

FIG. 179(*a*).

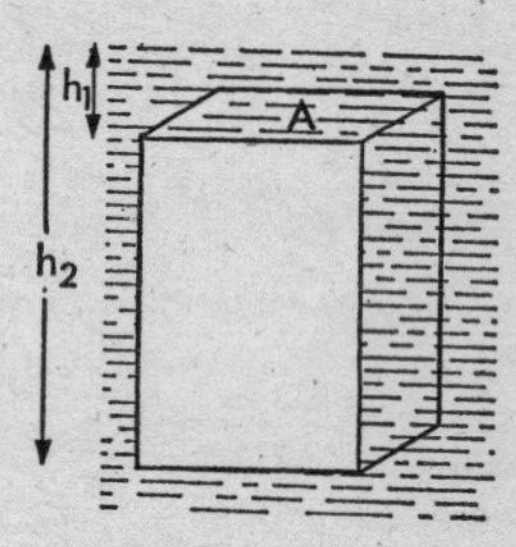

FIG. 179(*b*).

The pressure at the bottom of the body is

$$h\rho g \text{ N/m}^2.$$

Since pressure acts in all directions, an upwards thrust is applied to the body

$$= P \times A$$
$$= h\rho g A$$
$$= hA\rho g \text{ N.}$$

Now the volume of liquid which has been pushed out of the way $= h \times A$ m³.

And the mass of this liquid

$$= h \times A \times \rho \text{ kg.}$$

Its weight

$$= hA\rho g \text{ N.}$$

Thus the body receives an upthrust = weight of the displaced liquid.

If the body were completely submerged (Fig. 179(*b*)) the pressure on the top surface would be

$$h_1 \rho g \text{ N/m}^2,$$

and the pressure on the bottom surface would be

$$h_2 \rho g \text{ N/m}^2.$$

Thus the upthrust would be

$$(h_1 - h_2)A\rho g \text{ N}$$
$$= \text{weight of the displaced liquid.}$$

204. Forces "on" the liquid

If an object feels an upthrust from a liquid in which it is immersed we would expect there to be an equal and opposite force on the liquid itself. This can be easily checked by experiment, and it is worth checking the following examples.

Fig. 180(*a*) represents a body, hanging from a spring balance, which is about to be immersed in a liquid held in a container on another balance. When the body is placed in the liquid there is a reduction in the reading on the upper spring balance, A, and an increase in the reading of B (Fig. 180(*b*)). The two changes are equal.

Suppose the body has a volume V and the container has an area A.

The liquid level rises by $\frac{V}{A}$.

$\therefore$ There is an extra pressure $= \frac{V}{A}\rho g$

and an extra thrust on the base = pressure × area

$$= \frac{V}{A}\rho g \times A$$
$$= V\rho g$$
$$= \text{Weight of liquid displaced.}$$

If the container is very wide the increase in the level of the liquid will be only very small. The small increase in pressure will act on a very large area, however, and the extra thrust on the container will be equal to the weight of the displaced liquid.

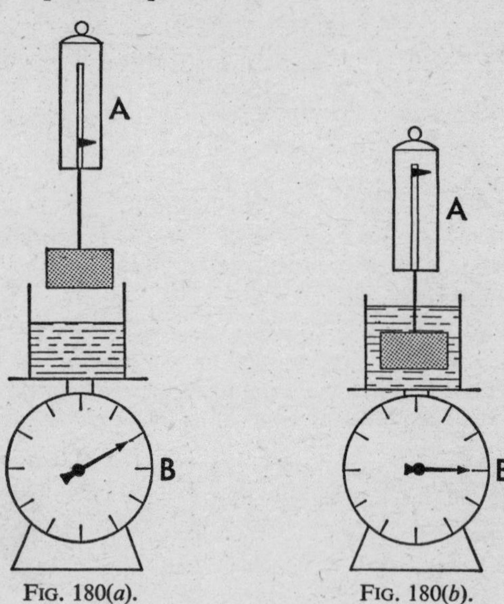

FIG. 180(*a*). FIG. 180(*b*).

205. Relative density and Archimedes, Principle

Archimedes' Principle gives us a quick method for finding the relative density of a liquid.

1. Place a can of water on a balance. Note the reading. W_1
2. Suspend an object in the water, making sure that it is not touching the bottom. Note the reading W_2
   $\therefore$ Upthrust from water $= W_2 - W_1$.
3. Replace the water by the liquid under test. Note the reading on the balance. W_3
4. Suspend the same object in the liquid. Note the new reading W_4
   $\therefore$ Upthrust from liquid $= W_4 - W_3$.

Now the upthrust on a body is equal to the weight of the liquid displaced.

∴ If the object had a volume V

$$\text{Relative density} = \frac{\text{Weight of a volume, } V\text{, of the liquid}}{\text{Weight of a volume, } V\text{, of water}}$$
$$= \frac{W_4 - W_3}{W_2 - W_1}.$$

206. Floating bodies

Suppose a body of weight W is placed in a liquid. Let w be the weight of the displaced liquid. There are three possibilities.

1. $W > w$ The body will sink.
2. $W < w$ The body will rise to the surface.
3. $W = w$ In this case the body and the liquid are in equilibrium.

In case 2 the body rises to the surface, but as soon as part of it emerges from the liquid it is no longer wholly immersed, and thus less liquid is displaced. Hence the upthrust, w, decreases. The weight, W, remains the same, so eventually we reach a state when the new upthrust $= W$. The body is now in equilibrium.

A body floats when the weight of the displaced fluid equals the weight of the body itself.

A body of density ρ_1 and volume V will, when placed in a liquid of density ρ_2, feel an upthrust of $V\rho_2 g$. Its own weight is $V\rho_1 g$.

Thus a solid body floats if $V\rho_1 g < V\rho_2 g$,

$$\text{i.e. if } \rho_1 < \rho_2.$$

A solid body floats if it has a density less than the liquid in which it has been immersed.

207. If a body immersed in a liquid is **hollow** instead of being solid, it displaces more of the liquid. Consequently the upward thrust, equal to the weight of displaced water, is greater than the weight of the solid. That is the reason why a ship, though built of iron, being hollow, displaces sufficient water for the upward thrust to be greater than the weight of the iron hull.

When the weight and displacement of the ship are known, it can be estimated what weight of cargo can be placed in the hold of the ship so that it does not sink so low as to be in danger of capsizing in high seas.

It should be noted that the density of salt water is greater than that of fresh water; the relative density varies between 1·025 and 1·028. Consequently, when a ship moves from fresh river-water to salt water, the weight of the water displaced is greater, the upward thrust increases, and the ship rises in the water. This is important in estimating the weight of the cargo which the ship can carry without sinking below the Plimsoll safety line, as it is called.

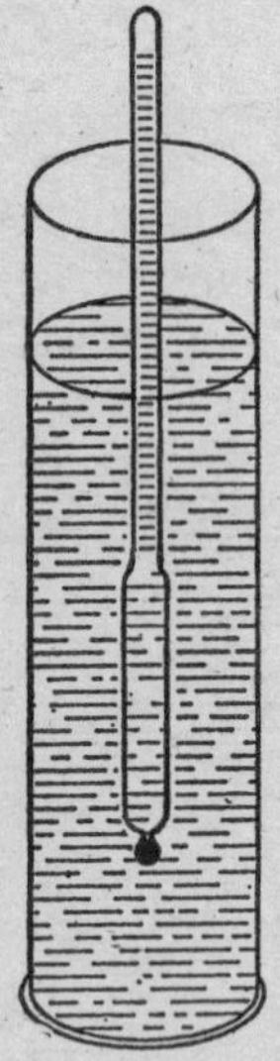

FIG. 181.

208. The hydrometer

The fact that a body floating in a liquid will rise or fall as the density of the liquid is increased or decreased suggests a convenient method of quickly determining the density of a liquid.

If a narrow hollow cylinder, weighted at the bottom to keep it vertical, be placed in liquids of known densities, the heights reached on the cylinder can be marked. Thus a graduated scale can be constructed. When the cylinder is placed in a liquid of unknown density, the mark reached on the scale by the level of the liquid will indicate its density.

This is the principle of the instrument known as the **Hydrometer.** The actual shape of the ordinary hydrometer is shown in Fig. 181. The cylinder with the graduated scale is long and slender, and below it is a bulb filled with air to give buoyancy. Below this is a smaller bulb containing mercury to keep the instrument vertical. The density is shown by the point reached on the graduated scale by the level of the liquid.

This instrument is used in industry for the rapid determination of the densities of milk, spirits, beer, etc. There are variations of the instrument for special purposes, but the general principle of their construction is the same.

209. Worked example

A rectangular block of wood 8 *cm by* 5 *cm by* 4 *cm and of relative density* 0·7 *floats in a liquid, and it is calculated that* 60 *cm*^{3} *of the wood are above the surface of the liquid. Find the relative density of the liquid.*

The volume of the wood = 160 cm^3.
Volume of liquid displaced = 160 − 60
= 100 cm^3.
Weight of wood = 0·7 × Weight of 160 cm^3 of water.

Since the wood displaces 100 cm^3 when floating,

Weight of wood = Weight of 100 cm^3 of liquid
= d × Weight of 100 cm^3 of water,

where d is the relative density of the liquid.

∴ 0·7 × Weight of 160 cm^3 of water = d × Weight of 100 cm^3 of water.

$$\therefore d = 0{\cdot}7 \times \frac{\text{Weight of 160 cm}^3\text{ of water}}{\text{Weight of 100 cm}^3\text{ of water}}$$

$$= 0{\cdot}7 \times \frac{160 \times 10^{-6} \times \rho \times g}{100 \times 10^{-6} \times \rho \times g}$$

$$= 0{\cdot}7 \times 1{\cdot}6$$

$$= 1{\cdot}12.$$

Exercise 29

(*The density of water is* 10^3 *kg/m*^{3}.)

1. Find the pressure at a point 50 m below the surface of the sea at a point where it has relative density 1·024.
2. A conical flask, 15 cm high, whose base has diameter 10 cm, is filled with a liquid of relative density 1·6. What is the thrust exerted by the liquid on the base?
3. A rectangular iron prism, 4 cm by 5 cm by 6 cm, is

placed in a bowl of mercury. $d_{iron} = 7{\cdot}2$; $d_{mercury} = 13{\cdot}6$. How much mercury is displaced?

4. 210 g of brass is suspended in water. $d_{brass} = 13{\cdot}6$. What is the upthrust on the brass? What would the upthrust be if the water were sea water ($d = 1{\cdot}024$)?

5. A cube of iron ($d = 7{\cdot}2$) has side 5 cm and is suspended by a wire in oil ($d = 0{\cdot}8$). What is the tension in the wire?

6. A solid, $d = 2{\cdot}4$, weighs 14.2 N. What is its weight in air?

7. After discharging 100 t of cargo, a ship has risen 10 cm in the water. If the water is fresh (i.e. $d = 1$) what is the average cross-sectional area of the ship at the water line?

8. A spring balance supporting a vessel full of water registers 9·2 N. A lump of metal weighing 12·6 N is suspended in the water, and does not touch the sides of the vessel. The new reading on the spring balance is 11 N. Find the relative density of the metal.

CHAPTER XIX

THE PRESSURE OF GASES

210. The pressure of the atmosphere

It is a curious circumstance that many people find it difficult to realise that gases have weight and can exert pressure. It is curious because we live in a mixture of gases, the atmosphere, and the pressure of the atmosphere is all-pervading, the whole of our body within and without being adapted to it. It is not easy to appreciate that each square centimetre of our body is subject to a force of 10 N, or about 1 kgf, so that, for example, if the palm of your hand has an area of about 100 cm^2, it is subjected to a force of 1 kN, or about 100 kgf. The fact that this is unnoticed, and that it involves no muscular effort, is due to the fact that a fluid, which includes both liquids and gases, **transmits pressure equally in all directions.** The pressure on the palm of the hand is counterbalanced by an equal pressure on the back of it. There is also an equal pressure within the human body; otherwise we would be squashed flat.

The atmosphere, being composed of gases, conforms to the law of fluids that pressure increases with depth below the surface. In this case we do not have a simple level for the surface. The density of the atmosphere gradually decreases as we go farther and farther away from the earth. A useful figure is 100 km, but observations on the passage of meteorites indicate that there is *some* gas as far away as 400 km.

At the earth's surface, therefore, there is a pressure caused by the effect of all the gas pressing down. The pressure decreases as we ascend, and use of this is made in altimeters.

This external change of pressure, if made rapidly, has harmful effects upon the human body, as the internal pressures do not so quickly adapt themselves to external changes. Aviators, and even mountain climbers, have to take special precautions to avoid harmful effects to the body resulting from quick changes in pressure.

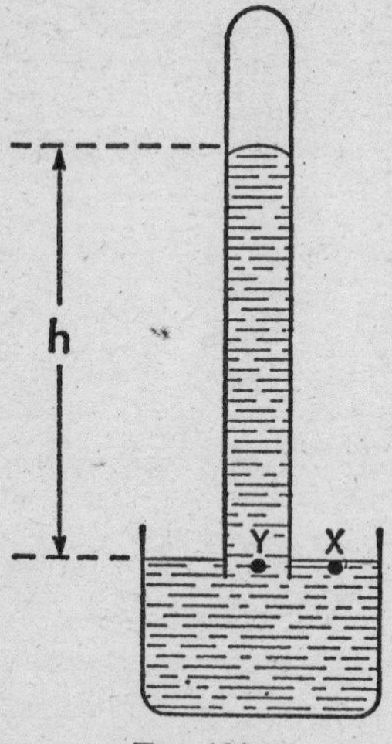

FIG. 182.

211. Measurement of air pressure

Many experiments can be employed to demonstrate the fact that the atmosphere exerts a pressure. It will be sufficient for our purposes to examine that devised by Torricelli, a pupil of Galileo.

Torricelli's experiment

A stout glass tube more than 80 cm long and sealed at one end is completely filled with mercury. A finger is placed over the open end and this end is placed beneath the surface of mercury in a bowl (Fig. 182). The finger is then withdrawn.

The level of mercury in the tube immediately falls to about 76 cm above the surface of the mercury in the bowl.

Let us consider the implications of this experiment.

The pressure at X = pressure of the atmosphere.
The pressure at X = pressure at Y (same level in same liquid).
The pressure at Y = pressure from the column of mercury.

$$= h\rho g \quad (\rho = \text{density of mercury})$$
$$= 0{\cdot}76 \times 13{\cdot}6 \times 10^3 \times 9{\cdot}8$$
$$= 100\,000 \text{ N/m}^2 \text{ (approx.)}.$$

If the pressure from the atmosphere rises the surface of the mercury in the bowl will fall a little and the column of mercury will rise until the pressure at Y is again equal to that exerted by the atmosphere.

N.B. (*a*) The height of the mercury column must be the vertical height.

(*b*) The cross-sectional area of the tube is irrelevant (see Fig. 173 on page 275).

Atmospheric pressure is frequently quoted as being 76 cm of mercury. This means that *the pressure is the same as that which would be produced by a column of mercury with 76 cm vertical height.*

If water were used instead of mercury the height would be

$$h = \frac{P}{g}$$
$$= \frac{100\,000}{1000 \times 9{\cdot}8}$$
$$= 10 \text{ m (approx.)}.$$

In meteorology the unit of pressure often used is the bar:

$$\mathbf{1\ bar = 100\,000\ N/m^2.}$$

Atmospheric pressure might be, e.g., 1020 millibars.

212. Total pressure in a liquid

In § 195 the pressure from a column of liquid was calculated. *The total pressure at any depth is the pressure from the liquid* ***plus*** *the pressure from, e.g., the air acting on the top of the liquid.*

213. The barometer

The pressure of the atmosphere is not a fixed quantity, but varies within certain limits. As stated previously, it varies with the height above sea-level. It also varies with the weather. Generally speaking, the pressure increases in fine weather and decreases in wet and stormy weather. To measure these changes in pressure the barometer ("pressure measurer") is used.

The instrument is essentially the same as that used in Torricelli's experiment. A scale is fixed by the side of the top of the mercury so that changes in the height of the mercury can be read.

Adjustments have been devised, as in the Fortin

barometer, to enable corrections to be made for the changes in the level of the mercury in the bowl. There is also a vernier attachment so that changes in height can be read to 0·1 mm.

The barometer is not only used to forecast changes in the weather, but it is essential in a large number of scientific experiments in which the atmospheric pressure is an important factor.

214. The aneroid barometer

The long column of mercury essential to the barometer is inconvenient for many purposes, and we know of no heavier suitable liquid of which a shorter column would be sufficient. When, therefore, no very great accuracy is necessary, the **aneroid** barometer is used. In this instrument no liquid is employed (aneroid = without liquid).

It consists of an enclosed cylinder, from which the air has been partly exhausted, and having its sides made of thin metal, usually corrugated to provide a larger surface. Owing to the partial exhaustion of air, the pressure on the external surface of the box is greater than on the internal. Changes in the atmospheric pressure produce slight expansions or contractions in the surface of the box. These changes, which are very small, are conveyed by rods and magnified by a system of small levers. Ultimately they are registered by a pointer on a circular dial scale.

215. Pumps

The pressure from the atmosphere is made use of in a number of pumps, one example of which, the Force Pump, is shown in Fig. 183.

When the piston P, which is air-tight, is raised the pressure at X falls, and the pressure from the atmosphere pushes water up the tube past the valve V_1.

When the piston is lowered, valve V_1 shuts, but the increased pressure opens V_2, and the water is forced out of the tube.

Raising P again results in valve V_1 opening and V_2

closing, so the operation can be maintained.

The reader should now examine the practical hydraulic lift in Fig. 177, page 280. Continuous operation of the small piston raises the load, and the larger piston can be lowered by opening the tap T.

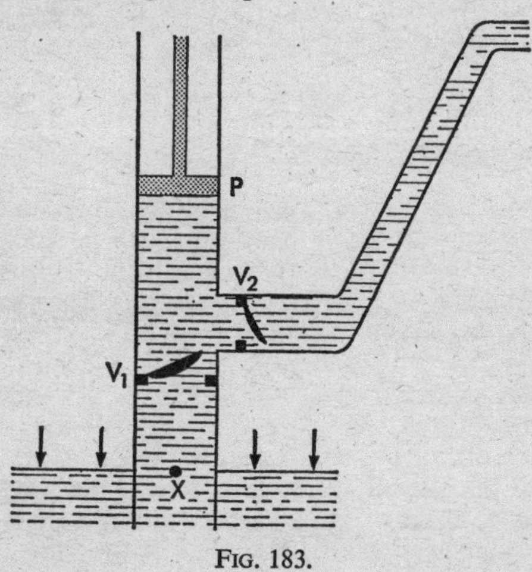

FIG. 183.

216. The siphon

Another useful device is the siphon, which is used in transferring liquid from a container at a high level to one at a low level. It is illustrated in Fig. 184. The tube ABC is filled with liquid and the ends are closed by the fingers. One end A is put below the surface of the upper container, while the end C is placed at a lower level. When the two ends are opened liquid flows from the higher level to the lower one.

The pressure at B can be calculated both for the left-hand side, relative to A, or for the right-hand side relative to C.

(a) Pressure at A = Pressure at $B + h_1 \rho g$.

$\therefore$ Pressure on left-hand side of B

= Pressure at $A - h_1 \rho g$.

Similarly

(b) Pressure on right-hand side of B

$$= \text{Pressure at } C - h_2\rho g.$$

Now pressure at A = pressure at C (atmospheric pressure).

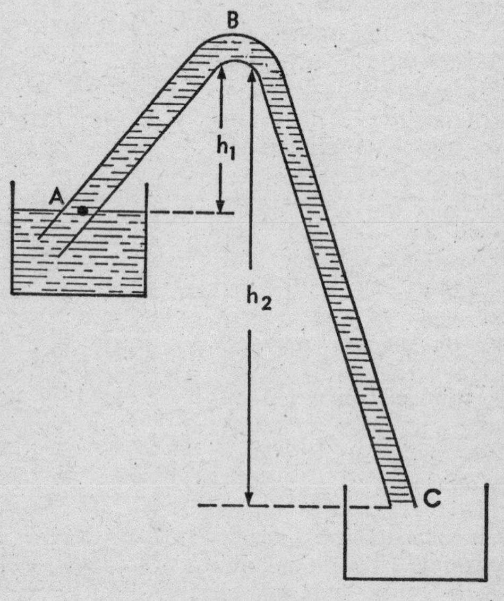

FIG. 184.

∴ Pressure on left-hand side of B is greater than the pressure on the right-hand side of B.

The pressure difference = $(h_2 - h_1)\rho g$.

It is this pressure difference which causes the liquid to flow along the tube until the upper vessel is emptied.

217. Boyle's Law

As has been stated in § 209, liquids are very slightly compressible, but gases change their volumes with changes in pressure and temperature. In this book, changes due to different pressures only will be examined.

As might be expected, the volumes of gases decrease when the pressure on them is increased, and conversely. The behaviour of gases under pressure was systematically examined by Robert Boyle, who published the results of his experiments in 1661.

One of the discoveries he announced is the principle known as Boyle's Law. This can be demonstrated by means of the apparatus shown in Fig. 185.

A small amount of dried air is passed through mercury into a glass tube *CD*, closed at one end and clamped to a vertical scale on which the volume of the air may be read.

This tube is connected by means of a stout rubber tube with **another vessel** ***AB*** containing mercury. The height of the mercury in *AB* is also registered on a vertical scale. The vessel is open, and the surface of the mercury is consequently subject to atmospheric pressure.

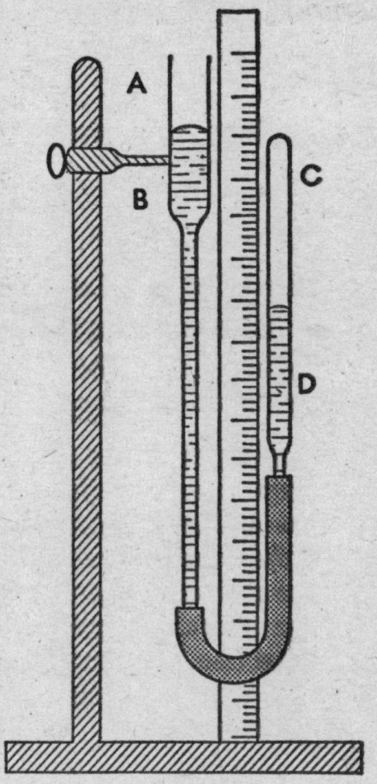

FIG. 185.

The vessel *AB* can be moved up and down vertically.

Thus, the surface of the mercury being at different heights, different pressures due to the mercury are transmitted through the rubber tube to the mercury in the vessel *CD*, and so to the air above the mercury.

As *AB* is raised, the pressure on the air in *CD* is increased and the volume of air decreased.

When *AB* is lowered, the pressure on the air in *CD* decreases and the volume of the air increases.

A number of experiments are made at different

pressures, and the corresponding volumes of the air are measured.

Let p be one of the pressures recorded.

Let v be the corresponding volume of air.

In each experiment the product $p \times v$ is calculated.

As a result of the experiments together with similar experiments on other gases it is found that *the product of* p *and* v *is constant.*

Thus $$p \times v = \textbf{constant.}$$

If k be the constant,

then $$pv = k$$

$$\text{and } p = \frac{k}{v}.$$

This is Boyle's Law

It can be expressed as follows:

Boyle's Law. *The volume of a mass of gas is inversely proportional to the pressure on it.*

It is important to remember that the temperature must be kept constant throughout the experiments, and the law would more accurately be expressed thus:

If the temperature remains constant the volume of a mass of gas is inversely proportional to the pressure on it.

The actual compression of the gas is accompanied by a slight rise in temperature, and for accuracy the reading should not be taken until the temperature has fallen to its previous level.

Exercise 30

The density of water is 1000 kg/m^3

Substance	Relative density (d)
Mercury . .	13·6
Glycerine . .	1·3
Sea water . .	1·024
Air . . .	0·0012

1. When the height of mercury in the barometer is 760 mm, what is the atmospheric pressure in millibars?
2. If the height of mercury is 758 mm, what would be the height if glycerine were used in the barometer?
3. What is the total pressure 50 m below the surface of a fresh-water lake when atmospheric pressure is 10^5 N/m^2?
4. What is the total pressure 50 m below the surface of the sea when the height of mercury in the barometer is 750 mm?
5. A mass of gas has volume 500 cm^3 when the pressure is 760 mm of mercury. What will its volume be when the pressure is: (1) 1250 mm of mercury; (2) 500 mm of mercury?
6. A glass cylinder contains 1 litre of air at a pressure of 2 atmospheres; air is slowly released until the pressure is 1·5 atmospheres. What fraction of the air is released and what volume does the released air occupy?
7. The atmospheric pressure at the base of a 1000-m high mountain is 760 mm of mercury. How much will the mercury level drop if the barometer is taken to the top of the mountain?

1. An air cylinder has a capacity of 0·5 m^3 and is initially filled with air at 25 MN/m^2. After use it is found that the pressure is 10 MN/m^2. What volume of air, measured at an atmospheric pressure of 10^5 N/m^2, has been used?

Revision Exercises

All answers will be given using approved SI units only, except for angles, which will be measured in degrees (°) rather than radians.

1. A wheelbarrow has wheels fixed 0·4 m from the end and 1·2 m from the tips of the handles. A mass of 80 kg rests at the front of the barrow. What vertical force applied at the handles will keep the barrow horizontal?

2. A man walks west for 5 km, north-east for 4 km and, finally, south for 3 km. If he had gone to his destination direct, in what direction should he have set off and how far would he have walked?

3. Two particles are projected simultaneously from the same point and subsequently move freely under gravity in the same vertical plane. Show that the velocity of one particle relative to the other is constant. Hence, or otherwise, show that the separation of the two particles is proportional to the time which has elapsed since projection.

4. Prove that if three forces are acting at a point and are in equilibrium each force is proportional to the sine of the angle between the other two.

A weight of 100 N hangs from two strings which are 7 m and 24 m long. They are attached to a horizontal ceiling, 25 m apart. What are the tensions in the strings?

5. Two particles rest on opposite sides of a rough double-inclined plane, connected by a string which runs over a pulley at the apex of the two planes. One particle weighs 11 N and rests on a 30° slope, the other is 2 N and rests on a 60° slope. What is the coefficient of friction (equal for both sides of the plane) if the system is in equilibrium?

6. A uniform circular lamina has radius 40 cm and has a hole of radius 5 cm cut in it. The centre of the hole is 5 cm from the centre of the lamina. Where is the lamina's centre of gravity?

7. A man wishes to lift a circular manhole lid by means of a ring diametrically opposite the hinge. The lid is

0·35 m in diameter and has mass 30 kg. What force must he apply?

8. A roller of 225 kg is pulled with its handle at 45° to the horizontal. If the coefficient of friction between the ground and roller is 1·3, what force does it need?

9. The roller of the last question is moved 6·6 m. How much work is done?

10. The work of the last question is accomplished in 1·5 min. What power was generated?

11. A uniform smooth sphere, of weight 20 N and radius 15 cm, rests on a smooth plane which is inclined at an angle of 30° to the horizontal. The sphere is held in equilibrium by a string of length 24 cm joining a point on the surface of the sphere to a point on the plane. Find the tension in the string and the force exerted by the sphere on the plane. (Cos $(A + B) = \cos A \cos B - \sin A \sin B$.)

12. A man of mass 75 kg is standing on one end of a stationary sledge of mass 225 kg. The sledge is on ice, assumed frictionless, and the man runs the length of the sledge, 2 m. How far does the sledge move?

13. Two uniform rough spheres are equal; their weight is W. They rest in contact on a rough horizontal plane. A force F acts on one of the spheres, along their line of centres and towards the other sphere. If the system is in equilibrium, what is the frictional force at each of the contacts? Show their directions by means of a diagram.

14. What pressure will hold a patch on to the inside of a hole in a boat if the hole is 50 cm below the surface and is of area 4 cm^2, and the patch has area 20 cm^2? The water is fresh.

15. A cat of mass 8 kg stands in a lift which rises with upward acceleration 7 m/s^2. What force does it exert on the floor? If each paw has area 8 cm^2, what pressure does the cat exert on the floor?

16. Two particles travel towards each other with velocities perpendicular and equal to 3 m/s and with equal masses of 5 μg. After colliding two particles emerge and continue in the original two directions, but they now have masses of 6 μg and 4 μg. What will their speeds be?

17. Wind, blowing at 9 km/h, hits a yacht's sail head

on. The sail has area 5 m^2 and the density of air is 1·2 kg/m^3. What is the thrust on the yacht's sail? The sail is at 60° to the direction of the ship's motion. What is the component of thrust in that direction?

18. A wooden cube ($d = 0{\cdot}8$) floats in water. What percentage of the depth of the wood is under water?

19. Two particles of mass 4 g and 3 g move towards each other at an angle of 45° at speeds of 12 m/s and 15 m/s respectively. After colliding they move as one particle. What is the velocity of the composite particle? (No momentum is lost in the collision.)

20. 750 t of water fall through 100 m each hour on to a turbine whose efficiency is 80 per cent. What power is generated?

ANSWERS

Answers to earlier exercises are given in fractional form where this is more convenient; answers given as decimals will be given to an accuracy warranted by the question.

p. 31 Exercise 1

1. (*a*) 0·064 m; (*b*) 9·6 kgf.
2. (*a*) 15 kgf; (*b*) 11·25 kgf; (*c*) 5 kgf.
3. 121·25 mm from the fulcrum.
4. 4·32 gf; (*b*) 216 cm from the fulcrum.
5. 14 gf.
6. 36·32 gf.
7. 22·3 kgf.
8. (*a*) 3 kgf; (*b*) 3·6 kgf.
9. 47·5 kgf.
10. 12 cm from 5 kgf weight.

p. 39 Exercise 2

1. $5\frac{5}{7}$; $4\frac{2}{7}$ kgf.
2. 19·75; 42·25 kgf.
3. $58\frac{1}{3}$; $71\frac{2}{3}$ kgf.
4. $F_D = 124$ kgf; $F_C = 93$ kgf.
5. $86\frac{2}{3}$; $113\frac{1}{3}$ kgf.
6. 5·25; 7·75 kgf.
7. $F_A = \frac{1}{3}$; $F_C = 1\frac{2}{3}$ tf.
8. 33·75; 71·25 kgf.

p. 48 Exercise 3

1. $4\frac{2}{7}$ kgf.
2. $\frac{1}{6}$ tf; $\frac{5}{6}$ tf.
3. $9\frac{1}{3}$ kgf; $1\frac{1}{3}$ kgf.
4. 5:7:2.
5. 39·5 kgf.
6. 28 kgf.
7. 1 m.
8. $\frac{2}{15}$ kgf/mm².
9. 120 kgf.
10. 10 kgf.

p. 64 Exercise 4

1. $1\frac{15}{16}$ m from *A*.
2. 1·85 m from *A*.
3. 0·75 m from *A*.
4. 8 kgf.
5. 0·566 m.
6. $\frac{7}{9}$ along the median from *A*.
7. 0·59 m from *E* along *EF*.
8. $4\frac{1}{3}$ m.
9. $1\frac{1}{2}$ m from *OA*; 1 m from *OB*.
10. 0·47 m.
11. $\left(5 + \frac{10\sqrt{3}}{3}\right)$ mm from *AB* —the c.g. of ΔPQR.
12. 50 mm from *AD*; 62·5 mm from *AB*.

p. 71 Exercise 5

1. 41·25.
2. On the axis and 0·2125 m from the lowest point of the hemisphere.
3. $16\frac{2}{3}$ mm from the base and on the axis.
4. 0·023 m from the centre.
5. 0·045; 0·035 m.
6. 7·7 mm, along the axis, from the base.
7. (i) neutral; (ii) unstable; (iii) stable; (iv) unstable except for very small displacements.

p. 87 Exercise 6

1. (*a*) 26; (*b*) 27·5; 36° 527′; (*c*) 12·2; (*d*) 8·72; 23° 25′; (*e*) 69° 35′; (*f*) $\theta = 78° 28'$; $\alpha = 44° 25'$
2. (1) 5·83 kgf; (2) 30° 59′.
3. (1) 20; (2) 19·32; (3) 18·48; (4) 18·12; (5) 17·32; (6) 14·14 kgf.
4. 11·6 kgf.
5. 60°.
6. 120°.
7. 102·6 kgf.
8. 18·03 kgf.
9. *BC* in 2:1 and *AC* in 3:1.

p. 95 Exercise 7

1. *OY*: 8 kgf; *OX*: 13.9 kgf.
2. *OX*: 10·88 kgf; *OY*: 5·07 kgf.
3. 4·59 kgf at 55° to *F*; 6·55 kgf.
4. 19·1 tf.
5. (1) 15·3 horizontally and 12·9 kgf vertically; (2) 216 gf horizontally and 125 gf vertically.
6. 94·0 kgf.
7. 10·35 kgf.
8. 26° 34′; 200 kgf.
9. 30·64 kgf.
10. 27·54 kgf.

p. 104 Exercise 8

1. 17·84 kgf.
2. 10·23 kgf; 81° 24′.
3. 10·15 kgf; 69° 49′.
4. 5·39 kgf at 23° 12′ to *AB*.
5. 5·35 kgf; 85° approx.
6. 6 units along the 5-unit string.
7. 21 gf; 57°.
8. $2\sqrt{3}$ tf; 30°.

p. 117 Exercise 9

1. $16\frac{2}{3}$ kgf; $13\frac{1}{3}$ kgf.
2. $F_3 = 18$ kgf; 2·5 m.
3. The second cord is at 30° with the vertical; 1·4 kgf; $1{\cdot}4\sqrt{3}$.
4. $F_R = 2{\cdot}46$ kgf; $T = 1{\cdot}72$ kgf.
5. $T = 2{\cdot}89$ kgf; $R = 5{\cdot}77$ kgf.
6. $F_2 = 24{\cdot}5$ kgf; $F_3 = 27{\cdot}3$ kgf.
7. $T_B = 2{\cdot}89$ kgf; $T_C = 8{\cdot}15$ kgf.
8. $BA = 6{\cdot}01$ tf; $AO = 12$ tf.

p. 122 Exercise 10

1. The calculated answer is 11·98 kgf; 75° 54′.
2. 89 kgf at 60° to F_1.
3. 3·2 kgf at 111° with F; the equilibriant is equal and opposite this.
4. 3·33; $\alpha = 70° 42'$.
5. 10·2 kgf; 38° 30′.
6. By Lami's theorem.

p. 134 Exercise 11

1. (*a*) 0·42; (*b*) 22° 47′.
2. 5·88 kgf.
3. 3·84 kgf.
4. 1·25 kgf.
5. 42·9 kg.
6. 1·875 kgf.
7. (1) 8 kgf; (2) 10·6; (3) 7·84 kgf.
9. 9·4 kgf.

p. 139 Exercise 12

1. (1) 0·404; (2) 1·87.
2. 17° 13′.
3. (1) 39° 37′; (2) 21·8 kgf.
4. 0·448.
5. 40 kgf.
6. 14·9 kgf.
7. 55·8 kgf.
8. 187 kgf.

p. 155 Exercise 13

1. 120 km/h.
2. 100 m.
3. 90 km/h.
4. $10{\cdot}792\,530 \times 10^8$ km/h.
5. 158·4 km/h.
6. 11 km/h.
7. In 2 s and 33·8 km.
8. 7·5; 5; 0.
9. 120; 60; 48; 36; 24; 12 km/h.
10. 115 m.

p. 165 Exercise 14

1. 40 m/s; 160 m.
2. 20 s.
3. 2·8 s; 15·4 m.
4. 0·25 m/s^2; 200 m.
5. $\frac{2}{3}$ m/s^2; $133\frac{1}{3}$ m.
6. 0·389 m/s^2; 14·4 s.
7. 2·06 km.
8. 38 m.
9. 28·125 km/h^2.
10. 17·36 m.
11. 18·5 s; 258·2 m.
12. 75 s.
13. 10 s.
14. 15·6 m; 5·6 s.

p. 170 Exercise 15

1.

Speed	0	4·9	9·8	14·7	19·6	24·5 m/s
Distance	0	1·225	4·9	11·05	19·6	30·625 m

2. 25·3 m.
3. 10·20 s; 510 m.
4. 122·5 m; 5 s.
5. 24·5 m/s; 30·65 m.
6. 14 m/s; 1·4 s.
7. 34·3 m.
8. 1·6 s; 4 s.
9. 195 m.
10. 15 m/s; 1 or 2 s.
11. 1·87 m/s²; 33·67.
12. 31·9 m; 5·1 s.

p. 181 Exercise 16

1. 0·4 m/s².
2. 96 N.
3. 2 kg.
4. 1·25 kN.
5. 0·35 m/s².
6. 7·81 kN.
7. 50 N.
8. 150 kN.
9. 1·2 kN.
10. 74 N.
11. 2×10^5 N.
12. 596 N.

p. 188 Exercise 17

1. 1·43 N; 2·8 m/s².
2. 240 N; 480 N.
3. 1·96 m/s².
4. 30·2 m.
5. 1225 mm/s.
6. 67·2 N; 1·4 m/s².
7. (*a*) 4·14 kgf; (*b*) 7 kgf; (*c*) 8·43 kgf.
8. 46·8 N.
9. 885 N.

p. 194 Exercise 18

1. (1) 12 500 kg m/s; (2) 25×10^{-5} kg m/s.
2. 60 cm/s.
3. 0·48 m/s.
4. 5 km/h.
5. 20 km/h.

p. 196 Exercise 19

1. 1·25 kN.
2. 145 kg m/s; 14·5 N.
3. 600 N.
4. 100 N.
5. 48 kg m/s; 2 m/s.
6. 60 N.

p. 207 Exercise 20

1. 159·6 km.
2. 9·68 N.
3. 468·75 N.
4. 0·531.
5. 14 m/s or 50·4 km/h.

p. 216 Exercise 21

1. (1) 360 J; (2) 25 kJ.
2. 7·5 kJ.
3. 80 kJ.
4. 1·25 kN.
5. 19 m.
6. 13·5 N.
7. 3·75 kN.
8. 981 J; 981 J.
9. 50 kN.
10. 1250 J; 9·6% of initial energy.
11. 504 N.
12. 245 kN.

p. 220 Exercise 22

1. 164 kW.
2. 1 kW.
3. 5·88 kW.
4. 114·5 W.
5. 630 N.
6. 27·5 MW.
7. 0·94 MW.
8. 2·6 kW.
9. 0·59 kW.
10. 10·625 kW.
11. 5 MN.
12. 67·5 W.

p. 241 Exercise 23

1. (*a*) $3\frac{1}{3}$; (*b*) 3·14; (*c*) 94%.
2. 71·4%.
3. 125 N.
4. 16 N; 37·5.
5. 0·65 or 65%.
6. 28 N.
7. 126; 18%.
8. (*a*) 294 N; (*b*) 96 N; (*c*) 75%.
9. $E = 0{\cdot}06L + 3{\cdot}5$.
10. $E = 0{\cdot}3L + 1{\cdot}1$.
11. $E = 0{\cdot}45L$.
12. $E = 0{\cdot}09L + 3{\cdot}5$.

p. 249 Exercise 24

1. (*a*) 17·44 m/s; 23° 26′. (*b*) 20·7 km/h; 21° 54′.
2. 15·3 km/h; 11° 18′ with the axis of the ship.
3. 5·6 knots; 67° 35′ N of W.
4. 1 h 9·3 min.
5. (By calculation) 195·96 km/h; 11° 32′.
6. 11° 51′ S. of E.; 206·8 km/h.
7. 1 h 46 min.
8. 36 km/h; 48 km/h.

p. 254 **Exercise 25**

1. 5 km/h; 36° 52′ S. of E.
2. 20·3 km/h; 47° 12′ N. of E.
3. 50 km/h; 53° 8′ E. of N.
4. 67 km/h; 63° 26′ N. of E.
5. 15 km/h; 36° 52′ W. of N.; 6 km.
6. 21·2 m/s; 266° 44′.
7. 28 knots; 27° S. of E.

p. 264 **Exercise 26**

1. $320\sqrt{3}$ m; 80 m.
2. 10·9 km.
3. 71° 5′; 18° 55′.
4. 250 m; $\pm$ 7·6 m.
5. 50 km.
6. 5·05; range = 125 m.
7. 14 m/s; 5 m.
8. 377·5 m; 510 m; 3·57 km.
9. 2·9 s; 200 m.
10. (*a*) 20 s; (*b*) 3500 m.
11. 91·83 m; 2·00%.

p. 269 **Exercise 27**

1. $11{\cdot}37 \times 10^3$ kg/m³; $d = 11{\cdot}37$.
2. 19·32 g.
3. 11%.
4. 1·06.
5. 3·26.

p. 272 **Exercise 28**

1. 114·45 N/m².
2. 0·588 MN/m².
3. 12 N.
4. 0·004 m².
5. 8·4 kN/m².

p. 287 **Exercise 29**

1. 0·50 MN/m².
2. 18·4 N/m².
3. 63·53 cm³.
4. 245 mN; 251 mN.
5. 7·84 N.
6. 24·34 N.
7. 1000 m³.
8. 7.

p. 296 **Exercise 30**

1. 1013 mbar.
2. 7930 mm.
3. $5{\cdot}9 \times 10^5$ N/m².
4. $6{\cdot}0 \times 10^5$ N/m².
5. (1) 304 cm³; (2) 760 cm³.
6. $\frac{1}{4}$; 0·5 l.
7. 88·2 mm.
8. 75 m³.

Revision Exercises

1. 261·6 N.
2. 4° 8′ W. of N.; 0·709 km.
4. 96 N; 28 N.
5. $5\sqrt{3} - 8$.
6. 0·08 cm from the centre along the axis.
7. 147 N.
8. 18·95 kN.
9. 82·3 kJ.
10. 0·91 kW.
11. **T** = 10·83; **N** = $\frac{5}{6}(\sqrt{3} + 5)$.
12. 0·5 m. (By cons. of momentum or, since no external force is acting, the centre of gravity of the system is static.)
13. $\frac{1}{2}F$ at each contact.
14. 981 N/m².
15. 22·4 N; 7 kN/m².
16. 3·75 m/s; 2·5 m/s.
17. 37·5 N; 18·75 N.
18. 80%.
19. 24·61 m/s making 18° 42′ with *OP* and on the side of *OP* opposite *OQ*.
20. 163·5 kW.

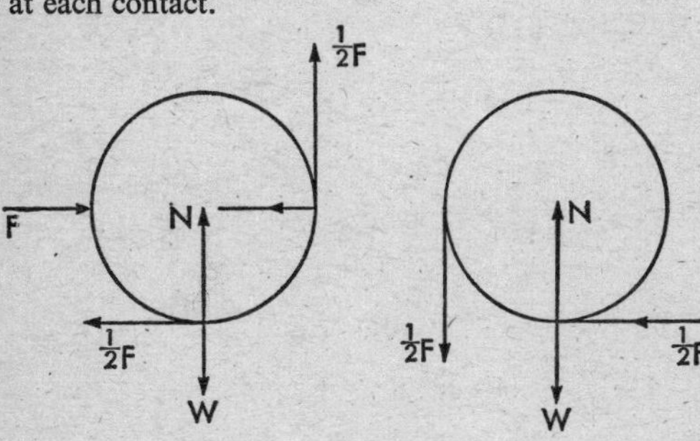

FIG. 186.

VALUES FOR SOME IMPERIAL UNITS IN SI UNITS

Length

1 in	25·4 mm
1 ft	304·8 mm
1 yd	0·91 m
1 mile	1·61 km

Velocity

1 mph	0·447 m/s

Energy

1 ft lbf	1·356 J

Power

1 hp	745·7 W

Mass

1 lb	0·454 kg
1 ton	1·016 t
1 ton	1016 kg

Density

1 lb/UK gal	99·7 kg/m^3
1 lb/ft^3	16·0 kg/m^3

Force

1 lbf	4·45 N

Pressure

1 lbf/in^2	6·89 kN/m^2

LOGARITHMS of numbers 100 to 549

Proportional Parts

	0	1	2	3	4	5	6	7	8	9	1	2	3	4	5	6	7	8	9
10	0000	0043	0086	0128	0170	0212	0253	0294	0334	0374	4	8	12	17	21	25	29	33	37
11	0414	0453	0492	0531	0569	0607	0645	0682	0719	0755	4	8	11	15	19	23	26	30	34
12	0792	0828	0864	0899	0934	0969	1004	1038	1072	1106	3	7	10	14	17	21	24	28	31
13	1139	1173	1206	1239	1271	1303	1335	1367	1399	1430	3	6	10	13	16	19	23	26	29
14	1461	1492	1523	1553	1584	1614	1644	1673	1703	1732	3	6	9	12	15	18	21	24	27
15	1761	1790	1818	1847	1875	1903	1931	1959	1987	2014	3	6	8	11	14	17	20	22	25
16	2041	2068	2095	2122	2148	2175	2201	2227	2253	2279	3	5	8	11	13	16	18	21	24
17	2304	2330	2355	2380	2405	2430	2455	2480	2504	2529	2	5	7	10	12	15	17	20	22
18	2553	2577	2601	2625	2648	2672	2695	2718	2742	2765	2	5	7	9	12	14	16	19	21
19	2788	2810	2833	2856	2878	2900	2923	2945	2967	2989	2	4	7	9	11	13	16	18	20
20	3010	3032	3054	3075	3096	3118	3139	3160	3181	3201	2	4	6	8	11	13	15	17	19
21	3222	3243	3263	3284	3304	3324	3345	3365	3385	3404	2	4	6	8	10	12	14	16	18
22	3424	3444	3464	3483	3502	3522	3541	3560	3579	3598	2	4	6	8	10	12	14	15	17
23	3617	3636	3655	3674	3692	3711	3729	3747	3766	3784	2	4	6	7	9	11	13	15	17
24	3802	3820	3838	3856	3874	3892	3909	3927	3945	3962	2	4	5	7	9	11	12	14	16
25	3979	3997	4014	4031	4048	4065	4082	4099	4116	4133	2	3	5	7	9	10	12	14	15
26	4150	4166	4183	4200	4216	4232	4249	4265	4281	4298	2	3	5	7	8	10	11	13	15
27	4314	4330	4346	4362	4378	4393	4409	4425	4440	4456	2	3	5	6	8	9	11	13	14
28	4472	4487	4502	4518	4533	4548	4564	4579	4594	4609	2	3	5	6	8	9	11	12	14
29	4624	4639	4654	4669	4683	4698	4713	4728	4742	4757	1	3	4	6	7	9	10	12	13
30	4771	4786	4800	4814	4829	4843	4857	4871	4886	4900	1	3	4	6	7	9	10	11	13
31	4914	4928	4942	4955	4969	4983	4997	5011	5024	5038	1	3	4	5	7	8	10	11	12
32	5051	5065	5079	5092	5105	5119	5132	5145	5159	5172	1	3	4	5	7	8	9	11	12
33	5185	5198	5211	5224	5237	5250	5263	5276	5289	5302	1	3	4	5	6	8	9	10	12
34	5315	5328	5340	5353	5366	5378	5391	5403	5416	5428	1	3	4	5	6	8	9	10	11
35	5441	5453	5465	5478	5490	5502	5514	5527	5539	5551	1	2	4	5	6	7	9	10	11
36	5563	5575	5587	5599	5611	5623	5635	5647	5658	5670	1	2	4	5	6	7	8	10	11
37	5632	5694	5705	5717	5729	5740	5752	5763	5775	5786	1	2	3	5	6	7	8	9	10
38	5798	5809	5821	5832	5843	5855	5866	5877	5888	5899	1	2	3	5	6	7	8	9	10
39	5911	5922	5933	5944	5955	5966	5977	5988	5999	6010	1	2	3	4	5	7	8	9	10
40	6021	6031	6042	6053	6064	6075	6085	6096	6107	6117	1	2	3	4	5	6	7	9	10
41	6128	6138	6149	6160	6170	6180	6191	6201	6212	6222	1	2	3	4	5	6	7	8	9
42	6232	6243	6253	6263	6274	6284	6294	6304	6314	6325	1	2	3	4	5	6	7	8	9
43	6335	6345	6355	6365	6375	6385	6395	6405	6415	6425	1	2	3	4	5	6	7	8	9
44	6435	6444	6454	6464	6474	6484	6493	6503	6513	6522	1	2	3	4	5	6	7	8	9
45	6532	6542	6551	6561	6571	6580	6590	6599	6609	6618	1	2	3	4	5	6	7	8	9
46	6628	6637	6646	6656	6665	6675	6684	6693	6702	6712	1	2	3	4	5	6	7	7	8
47	6721	6730	6739	6749	6758	6767	6776	6785	6794	6803	1	2	3	4	5	5	6	7	8
48	6812	6821	6830	6839	6848	6857	6866	6875	6884	6893	1	2	3	4	4	5	6	7	8
49	6902	6911	6920	6928	6937	6946	6955	6964	6972	6981	1	2	3	4	4	5	6	7	8
50	6990	6998	7007	7016	7024	7033	7042	7050	7059	7067	1	2	3	3	4	5	6	7	8
51	7076	7084	7093	7101	7110	7118	7126	7135	7143	7152	1	2	3	3	4	5	6	7	8
52	7160	7168	7177	7185	7193	7202	7210	7218	7226	7235	1	2	2	3	4	5	6	7	7
53	7243	7251	7259	7267	7275	7284	7292	7300	7308	7316	1	2	2	3	4	5	6	6	7
54	7324	7332	7340	7348	7356	7364	7372	7380	7388	7396	1	2	2	3	4	5	6	6	7
	0	1	2	3	4	5	6	7	8	9	1	2	3	4	5	6	7	8	9

LOGARITHMS of numbers 550 to 999

	0	1	2	3	4	5	6	7	8	9	Proportional Parts 1	2	3	4	5	6	7	8	9
55	7404	7412	7419	7427	7435	7443	7451	7459	7466	7474	1	2	2	3	4	5	5	6	7
56	7482	7490	7497	7505	7513	7520	7528	7536	7543	7551	1	2	2	3	4	5	5	6	7
57	7559	7566	7574	7582	7589	7597	7604	7612	7619	7627	1	2	2	3	4	5	5	6	7
58	7634	7642	7649	7657	7664	7672	7679	7686	7694	7701	1	1	2	3	4	4	5	6	7
59	7709	7716	7723	7731	7738	7745	7752	7760	7767	7774	1	1	2	3	4	4	5	6	7
60	7782	7789	7796	7803	7810	7818	7825	7832	7839	7846	1	1	2	3	4	4	5	6	6
61	7853	7860	7868	7875	7882	7889	7896	7903	7910	7917	1	1	2	3	4	4	5	6	6
62	7924	7931	7938	7945	7952	7959	7966	7973	7980	7987	1	1	2	3	3	4	5	6	6
63	7993	8000	8007	8014	8021	8028	8035	8041	8048	8055	1	1	2	3	3	4	5	6	6
64	8062	8069	8075	8082	8089	8096	8102	8109	8116	8122	1	1	2	3	3	4	5	5	6
65	8129	8136	8142	8149	8156	8162	8169	8176	8182	8189	1	1	2	3	3	4	5	5	6
66	8195	8202	8209	8215	8222	8228	8235	8241	8248	8254	1	1	2	3	3	4	5	5	6
67	8261	8267	8274	8280	8287	8293	8299	8306	8312	8319	1	1	2	3	3	4	4	5	6
68	8325	8331	8338	8344	8351	8357	8363	8370	8376	8382	1	1	2	3	3	4	4	5	6
69	8388	8395	8401	8407	8414	8420	8426	8432	8439	8445	1	1	2	3	3	4	4	5	6
70	8451	8457	8463	8470	8476	8482	8488	8494	8500	8506	1	1	2	2	3	4	4	5	6
71	8513	8519	8525	8531	8537	8543	8549	8555	8561	8567	1	1	2	2	3	4	4	5	5
72	8573	8579	8585	8591	8597	8603	8609	8615	8621	8627	1	1	2	2	3	4	4	5	5
73	8633	8639	8645	8651	8657	8663	8669	8675	8681	8686	1	1	2	2	3	4	4	5	5
74	8692	8698	8704	8710	8716	8722	8727	8733	8739	8745	1	1	2	2	3	4	4	5	5
75	8751	8756	8762	8768	8774	8779	8785	8791	8797	8802	1	1	2	2	3	3	4	5	5
76	8808	8814	8820	8825	8831	8837	8842	8848	8854	8859	1	1	2	2	3	3	4	5	5
77	8865	8871	8876	8882	8887	8893	8899	8904	8910	8915	1	1	2	2	3	3	4	4	5
78	8921	8927	8932	8938	8943	8949	8954	8960	8965	8971	1	1	2	2	3	3	4	4	5
79	8976	8982	8987	8993	8998	9004	9009	9015	9020	9025	1	1	2	2	3	3	4	4	5
80	9031	9036	9042	9047	9053	9058	9063	9069	9074	9079	1	1	2	2	3	3	4	4	5
81	9085	9090	9096	9101	9106	9112	9117	9122	9128	9133	1	1	2	2	3	3	4	4	5
82	9138	9143	9149	9154	9159	9165	9170	9175	9180	9186	1	1	2	2	3	3	4	4	5
83	9191	9196	9201	9206	9212	9217	9222	9227	9232	9238	1	1	2	2	3	3	4	4	5
84	9243	9248	9253	9258	9263	9269	9274	9279	9284	9289	1	1	2	2	3	3	4	4	5
85	9294	9299	9304	9309	9315	9320	9325	9330	9335	9340	1	1	2	2	3	3	4	4	5
86	9345	9350	9355	9360	9365	9370	9375	9380	9385	9390	1	1	2	2	3	3	4	4	5
87	9395	9400	9405	9410	9415	9420	9425	9430	9435	9440	0	1	1	2	2	3	3	4	4
88	9445	9450	9455	9460	9465	9469	9474	9479	9484	9489	0	1	1	2	2	3	3	4	4
89	9494	9499	9504	9509	9513	9518	9523	9528	9533	9538	0	1	1	2	2	3	3	4	4
90	9542	9547	9552	9557	9562	9566	9571	9576	9581	9586	0	1	1	2	2	3	3	4	4
91	9590	9595	9600	9605	9609	9614	9619	9624	9628	9633	0	1	1	2	2	3	3	4	4
92	9638	9643	9647	9652	9657	9661	9666	9671	9675	9680	0	1	1	2	2	3	3	4	4
93	9685	9689	9694	9699	9703	9708	9713	9717	9722	9727	0	1	1	2	2	3	3	4	4
94	9731	9736	9741	9745	9750	9754	9759	9764	9768	9773	0	1	1	2	2	3	3	4	4
95	9777	9782	9786	9791	9795	9800	9805	9809	9814	9818	0	1	1	2	2	3	3	4	4
96	9823	9827	9832	9836	9841	9845	9850	9854	9859	9863	0	1	1	2	2	3	3	4	4
97	9868	9872	9877	9881	9886	9890	9894	9899	9903	9908	0	1	1	2	2	3	3	4	4
98	9912	9917	9921	9926	9930	9934	9939	9943	9948	9952	0	1	1	2	2	3	3	4	4
99	9956	9961	9965	9969	9974	9978	9983	9987	9991	9996	0	1	1	2	2	3	3	4	4
	0	1	2	3	4	5	6	7	8	9	1	2	3	4	5	6	7	8	9

ANTI-LOGARITHMS

	0	1	2	3	4	5	6	7	8	9	Proportional Parts 1	2	3	4	5	6	7	8	9
·00	1000	1002	1005	1007	1009	1012	1014	1016	1019	1021	0	0	1	1	1	1	2	2	2
·01	1023	1026	1028	1030	1033	1035	1038	1040	1042	1045	0	0	1	1	1	1	2	2	2
·02	1047	1050	1052	1054	1057	1059	1062	1064	1067	1069	0	0	1	1	1	1	2	2	2
·03	1072	1074	1076	1079	1081	1084	1086	1089	1091	1094	0	0	1	1	1	1	2	2	2
·04	1096	1099	1102	1104	1107	1109	1112	1114	1117	1119	0	1	1	1	1	2	2	2	2
·05	1122	1125	1127	1130	1132	1135	1138	1140	1143	1146	0	1	1	1	1	2	2	2	2
·06	1148	1151	1153	1156	1159	1161	1164	1167	1169	1172	0	1	1	1	1	2	2	2	2
·07	1175	1178	1180	1183	1186	1189	1191	1194	1197	1199	0	1	1	1	1	2	2	2	2
·08	1202	1205	1208	1211	1213	1216	1219	1222	1225	1227	0	1	1	1	1	2	2	2	3
·09	1230	1233	1236	1239	1242	1245	1247	1250	1253	1256	0	1	1	1	1	2	2	2	3
·10	1259	1262	1265	1268	1271	1274	1276	1279	1282	1285	0	1	1	1	1	2	2	2	3
·11	1288	1291	1294	1297	1300	1303	1306	1309	1312	1315	0	1	1	1	2	2	2	2	3
·12	1318	1321	1324	1327	1330	1334	1337	1340	1343	1346	0	1	1	1	2	2	2	2	3
·13	1349	1352	1355	1358	1361	1365	1368	1371	1374	1377	0	1	1	1	2	2	2	2	3
·14	1380	1384	1387	1390	1393	1396	1400	1403	1406	1409	0	1	1	1	2	2	2	3	3
·15	1413	1416	1419	1422	1426	1429	1432	1435	1439	1442	0	1	1	1	2	2	2	3	3
·16	1445	1449	1452	1455	1459	1462	1466	1469	1472	1476	0	1	1	1	2	2	2	3	3
·17	1479	1483	1486	1489	1493	1496	1500	1503	1507	1510	0	1	1	1	2	2	2	3	3
·18	1514	1517	1521	1524	1528	1531	1535	1538	1542	1545	0	1	1	1	2	2	2	3	3
·19	1549	1552	1556	1560	1563	1567	1570	1574	1578	1581	0	1	1	1	2	2	3	3	3
·20	1585	1589	1592	1596	1600	1603	1607	1611	1614	1618	0	1	1	1	2	2	3	3	3
·21	1622	1626	1629	1633	1637	1641	1644	1648	1652	1656	0	1	1	2	2	2	3	3	3
·22	1660	1663	1667	1671	1675	1679	1683	1687	1690	1694	0	1	1	2	2	2	3	3	3
·23	1698	1702	1706	1710	1714	1718	1722	1726	1730	1734	0	1	1	2	2	2	3	3	4
·24	1738	1742	1746	1750	1754	1758	1762	1766	1770	1774	0	1	1	2	2	2	3	3	4
·25	1778	1782	1786	1791	1795	1799	1803	1807	1811	1816	0	1	1	2	2	3	3	3	4
·26	1820	1824	1828	1832	1837	1841	1845	1849	1854	1858	0	1	1	2	2	3	3	3	4
·27	1862	1866	1871	1875	1879	1884	1888	1892	1897	1901	0	1	1	2	2	3	3	3	4
·28	1905	1910	1914	1919	1923	1928	1932	1936	1941	1945	0	1	1	2	2	3	3	4	4
·29	1950	1954	1959	1963	1968	1972	1977	1982	1986	1991	0	1	1	2	2	3	3	4	4
·30	1995	2000	2004	2009	2014	2018	2023	2028	2032	2037	0	1	1	2	2	3	3	4	4
·31	2042	2046	2051	2056	2061	2065	2070	2075	2080	2084	0	1	1	2	2	3	3	4	4
·32	2089	2094	2099	2104	2109	2113	2118	2123	2128	2133	0	1	1	2	2	3	3	4	4
·33	2138	2143	2148	2153	2158	2163	2168	2173	2178	2183	0	1	1	2	2	3	3	4	4
·34	2188	2193	2198	2203	2208	2213	2218	2223	2228	2234	1	1	2	2	3	3	4	4	5
·35	2239	2244	2249	2254	2259	2265	2270	2275	2280	2286	1	1	2	2	3	3	4	4	5
·36	2291	2296	2301	2307	2312	2317	2323	2328	2333	2339	1	1	2	2	3	3	4	4	5
·37	2344	2350	2355	2360	2366	2371	2377	2382	2388	2393	1	1	2	2	3	3	4	4	5
·38	2399	2404	2410	2415	2421	2427	2432	2438	2443	2449	1	1	2	2	3	3	4	4	5
·39	2455	2460	2466	2472	2477	2483	2489	2495	2500	2506	1	1	2	2	3	3	4	5	5
·40	2512	2518	2523	2529	2535	2541	2547	2553	2559	2564	1	1	2	2	3	3	4	5	5
·41	2570	2576	2582	2588	2594	2600	2606	2612	2618	2624	1	1	2	2	3	4	4	5	5
·42	2630	2636	2642	2648	2655	2661	2667	2673	2679	2685	1	1	2	2	3	4	4	5	6
·43	2692	2698	2704	2710	2716	2723	2729	2735	2742	2748	1	1	2	2	3	4	4	5	6
·44	2754	2761	2767	2773	2780	2786	2793	2799	2805	2812	1	1	2	3	3	4	4	5	6
·45	2818	2825	2831	2838	2844	2851	2858	2864	2871	2877	1	1	2	3	3	4	5	5	6
·46	2884	2891	2897	2904	2911	2917	2924	2931	2938	2944	1	1	2	3	3	4	5	5	6
·47	2951	2958	2965	2972	2979	2985	2992	2999	3006	3013	1	1	2	3	3	4	5	6	6
·48	3020	3027	3034	3041	3048	3055	3062	3069	3076	3083	1	1	2	3	4	4	5	6	6
·49	3090	3097	3105	3112	3119	3126	3133	3141	3148	3155	1	1	2	3	4	4	5	6	7
	0	1	2	3	4	5	6	7	8	9	1	2	3	4	5	6	7	8	9

ANTI-LOGARITHMS

	0	1	2	3	4	5	6	7	8	9	Proportional Parts 1	2	3	4	5	6	7	8	9
·50	3162	3170	3177	3184	3192	3199	3206	3214	3221	3228	1	1	2	3	4	4	5	6	7
·51	3236	3243	3251	3258	3266	3273	3281	3289	3296	3304	1	2	2	3	4	5	5	6	7
·52	3311	3319	3327	3334	3342	3350	3357	3365	3373	3381	1	2	2	3	4	5	5	6	7
·53	3388	3396	3404	3412	3420	3428	3436	3443	3451	3459	1	2	2	3	4	5	6	6	7
·54	3467	3475	3483	3491	3499	3508	3516	3524	3532	3540	1	2	2	3	4	5	6	6	7
·55	3548	3556	3565	3573	3581	3589	3597	3606	3614	3622	1	2	2	3	4	5	6	7	7
·56	3631	3639	3648	3656	3664	3673	3681	3690	3698	3707	1	2	3	3	4	5	6	7	8
·57	3715	3724	3733	3741	3750	3758	3767	3776	3784	3793	1	2	3	3	4	5	6	7	8
·58	3802	3811	3819	3828	3837	3846	3855	3864	3873	3882	1	2	3	4	4	5	6	7	8
·59	3890	3899	3908	3917	3926	3936	3945	3954	3963	3972	1	2	3	4	5	5	6	7	8
·60	3981	3990	3999	4009	4018	4027	4036	4046	4055	4064	1	2	3	4	5	6	7	7	8
·61	4074	4083	4093	4102	4111	4121	4130	4140	4150	4159	1	2	3	4	5	6	7	8	9
·62	4169	4178	4188	4198	4207	4217	4227	4236	4246	4256	1	2	3	4	5	6	7	8	9
·63	4266	4276	4285	4295	4305	4315	4325	4335	4345	4355	1	2	3	4	5	6	7	8	9
·64	4365	4375	4385	4395	4406	4416	4426	4436	4446	4457	1	2	3	4	5	6	7	8	9
·65	4467	4477	4487	4498	4508	4519	4529	4539	4550	4560	1	2	3	4	5	6	7	8	9
·66	4571	4581	4592	4603	4613	4624	4634	4645	4656	4667	1	2	3	4	5	6	7	8	10
·67	4677	4688	4699	4710	4721	4732	4742	4753	4764	4775	1	2	3	4	5	7	8	9	10
·68	4786	4797	4808	4819	4831	4842	4853	4864	4875	4887	1	2	3	4	6	7	8	9	10
·69	4898	4909	4920	4932	4943	4955	4966	4977	4989	5000	1	2	3	5	6	7	8	9	10
·70	5012	5023	5035	5047	5058	5070	5082	5093	5105	5117	1	2	4	5	6	7	8	9	11
·71	5129	5140	5152	5164	5176	5188	5200	5212	5224	5236	1	2	4	5	6	7	8	10	11
·72	5248	5260	5272	5284	5297	5309	5321	5333	5346	5358	1	2	4	5	6	7	9	10	11
·73	5370	5383	5395	5408	5420	5433	5445	5458	5470	5483	1	3	4	5	6	8	9	10	11
·74	5495	5508	5521	5534	5546	5559	5572	5585	5598	5610	1	3	4	5	6	8	9	10	12
·75	5623	5636	5649	5662	5675	5689	5702	5715	5728	5741	1	3	4	5	7	8	9	10	12
·76	5754	5768	5781	5794	5808	5821	5834	5848	5861	5875	1	3	4	5	7	8	9	11	12
·77	5888	5902	5916	5929	5943	5957	5970	5984	5998	6012	1	3	4	6	7	8	10	11	12
·78	6026	6039	6053	6067	6081	6095	6109	6124	6138	6152	1	3	4	6	7	8	10	11	13
·79	6166	6180	6194	6209	6223	6237	6252	6266	6281	6295	1	3	4	6	7	9	10	12	13
·80	6310	6324	6339	6353	6368	6383	6397	6412	6427	6442	1	3	4	6	7	9	10	12	13
·81	6457	6471	6486	6501	6516	6531	6546	6561	6577	6592	2	3	5	6	8	9	11	12	14
·82	6607	6622	6637	6653	6668	6683	6699	6714	6730	6745	2	3	5	6	8	9	11	12	14
·83	6761	6776	6792	6808	6823	6839	6855	6871	6887	6902	2	3	5	6	8	9	11	13	14
·84	6918	6934	6950	6966	6982	6998	7015	7031	7047	7063	2	3	5	6	8	10	11	13	14
·85	7079	7096	7112	7129	7145	7161	7178	7194	7211	7228	2	3	5	7	8	10	12	13	15
·86	7244	7261	7278	7295	7311	7328	7345	7362	7379	7396	2	3	5	7	8	10	12	14	15
·87	7413	7430	7447	7464	7482	7499	7516	7534	7551	7568	2	3	5	7	9	10	12	14	16
·88	7586	7603	7621	7638	7656	7674	7691	7709	7727	7745	2	4	5	7	9	11	12	14	16
·89	7762	7780	7798	7816	7834	7852	7870	7889	7907	7925	2	4	5	7	9	11	13	14	16
·90	7943	7962	7980	7998	8017	8035	8054	8072	8091	8110	2	4	6	7	9	11	13	15	17
·91	8128	8147	8166	8185	8204	8222	8241	8260	8279	8299	2	4	6	8	10	11	13	15	17
·92	8318	8337	8356	8375	8395	8414	8433	8453	8472	8492	2	4	6	8	10	12	14	15	17
·93	8511	8531	8551	8570	8590	8610	8630	8650	8670	8690	2	4	6	8	10	12	14	16	18
·94	8710	8730	8750	8770	8790	8810	8831	8851	8872	8892	2	4	6	8	10	12	14	16	18
·95	8913	8933	8954	8974	8995	9016	9036	9057	9078	9099	2	4	6	8	10	12	14	17	19
·96	9120	9141	9162	9183	9204	9226	9247	9268	9290	9311	2	4	6	9	11	13	15	17	19
·97	9333	9354	9376	9397	9419	9441	9462	9484	9506	9528	2	4	7	9	11	13	15	17	20
·98	9550	9572	9594	9616	9638	9661	9683	9705	9727	9750	2	4	7	9	11	13	16	18	20
·99	9772	9795	9817	9840	9863	9886	9908	9931	9954	9977	2	5	7	9	11	14	16	18	21
	0	1	2	3	4	5	6	7	8	9	1	2	3	4	5	6	7	8	9

NATURAL SINES

	0′	6′	12′	18′	24′	30′	36′	42′	48′	54′	Proportional Parts 1′	2′	3′	4′	5′
0°	0·0000	·0017	0035	0052	0070	0087	·0105	0122	·0140	·0157	3	6	9	12	15
1	0 0175	0192	0209	0227	0244	0262	0279	·0297	0314	·0332	3	6	9	12	15
2	0 0349	0366	0384	0401	0419	0436	0454	·0471	·0489	·0506	3	6	9	12	15
3	0 0523	0541	0558	0576	0593	·0610	·0628	0645	0663	·0680	3	6	9	12	15
4	0 0698	·0715	·0732	·0750	0767	·0785	·0802	0819	0837	·0854	3	6	9	12	14
5	0 0872	0889	·0906	0924	0941	0958	·0976	0993	1011	·1028	3	6	9	12	14
6	0 1045	1063	·1080	1097	·1115	·1132	1149	1167	·1184	·1201	3	6	9	12	14
7	0 1219	1236	·1253	1271	1288	1305	1323	·1340	1357	·1374	3	6	9	12	14
8	0 1392	1409	1426	1444	1461	·1478	·1495	·1513	1530	·1547	3	6	9	11	14
9	0 1564	1582	1599	1616	1633	·1650	·1668	1685	·1702	·1719	3	6	9	11	14
10	0 1736	1754	·1771	1788	1805	·1822	1840	·1857	1874	·1891	3	6	9	11	14
11	0 1908	1925	1942	·1959	·1977	1994	2011	2028	2045	2062	3	6	9	11	14
12	0·2079	·2096	2113	2130	·2147	2164	2181	·2198	2215	2232	3	6	9	11	14
13	0·2250	·2267	2284	2300	2317	·2334	·2351	·2368	2385	2402	3	6	8	11	14
14	0·2419	·2436	·2453	2470	2487	·2504	·2521	2538	·2554	·2571	3	6	8	11	14
15	0·2588	·2605	·2622	·2639	2656	2672	2689	2706	·2723	·2740	3	6	8	11	14
16	0 2756	·2773	2790	2807	2823	·2840	2857	·2874	2890	2907	3	6	8	11	14
17	0·2924	2940	2957	2974	2990	3007	3024	·3040	3057	3074	3	6	8	11	14
18	0·3090	·3107	·3123	3140	3156	·3173	·3190	·3206	·3223	·3239	3	6	8	11	14
19	0·3256	·3272	·3289	3305	3322	3338	3355	·3371	·3387	·3404	3	5	8	11	14
20	0 3420	·3437	·3453	3469	3486	3502	·3518	·3535	·3551	3567	3	5	8	11	14
21	0 3584	·3600	·3616	3633	3649	3665	·3681	·3697	3714	·3730	3	5	8	11	14
22	0 3746	·3762	·3778	3795	·3811	·3827	3843	·3859	3875	3891	3	5	8	11	13
23	0 3907	·3923	·3939	3955	3971	3987	4003	·4019	4035	4051	3	5	8	11	13
24	0·4067	·4083	·4099	4115	·4131	4147	4163	·4179	·4195	·4210	3	5	8	11	13
25	0·4226	·4242	4258	4274	4289	·4305	4321	·4337	·4352	4368	3	5	8	11	13
26	0 4384	·4399	4415	·4431	4446	4462	4478	4493	·4509	·4524	3	5	8	10	13
27	0·4540	·4555	·4571	·4586	4602	·4617	·4633	4648	·4664	·4679	3	5	8	10	13
28	0·4695	·4710	·4726	4741	4756	4772	4787	4802	·4818	4833	3	5	8	10	13
29	0 4848	·4863	·4879	·4894	·4909	4924	·4939	·4955	·4970	·4985	3	5	8	10	13
30	0 5000	·5015	·5030	·5045	·5060	·5075	·5090	·5105	·5120	·5135	2	5	8	10	12
31	0·5150	·5165	·5180	·5195	5210	·5225	5240	·5255	5270	·5284	2	5	7	10	12
32	0·5299	·5314	5329	·5344	·5358	·5373	5388	5402	·5417	·5432	2	5	7	10	12
33	0 5446	·5461	·5476	·5490	5505	5519	·5534	5548	·5563	·5577	2	5	7	10	12
34	0·5592	·5606	·5621	·5635	·5650	5664	·5678	5693	·5707	·5721	2	5	7	10	12
35	0·5736	·5750	5764	·5779	·5793	5807	·5821	5835	5850	·5864	2	5	7	9	12
36	0·5878	·5892	·5906	·5920	5934	5948	·5962	·5976	5990	6004	2	5	7	9	12
37	0·6018	·6032	·6046	·6060	·6074	6088	·6101	·6115	·6129	6143	2	5	7	9	12
38	0·6157	·6170	·6184	·6198	6211	·6225	·6239	·6252	·6266	·6280	2	5	7	9	11
39	0·6293	·6307	·6320	·6334	6347	·6361	·6374	·6388	6401	·6414	2	4	7	9	11
40	0·6428	·6441	6455	·6468	6481	·6494	·6508	·6521	·6534	·6547	2	4	7	9	11
41	0·6561	·6574	·6587	·6600	6613	·6626	·6639	·6652	·6665	·6678	2	4	6	9	11
42	0 6691	·6704	·6717	·6730	·6743	·6756	·6769	·6782	·6794	·6807	2	4	6	9	11
43	0·6820	·6833	·6845	·6858	·6871	·6884	·6896	·6909	·6921	6934	2	4	6	8	11
44	0·6947	·6959	·6972	·6984	·6997	·7009	·7022	·7034	·7046	·7059	2	4	6	8	10
	0′	6′	12′	18′	24′	30′	36′	42′	48′	54′	1′	2′	3′	4′	5′

NATURAL SINES

	0′	6′	12′	18′	24′	30′	36′	42′	48′	54′	Proportional Parts 1′	2′	3′	4′	5′
45°	0.7071	7083	7096	7108	7120	7133	7145	7157	7169	7181	2	4	6	8	10
46	0.7193	7206	7218	7230	7242	7254	7266	7278	7290	7302	2	4	6	8	10
47	0.7314	7325	7337	7349	7361	7373	7385	7396	7408	7420	2	4	6	8	10
48	0.7431	7443	7455	7466	7478	7490	7501	7513	7524	7536	2	4	6	8	10
49	0.7547	7559	7570	7581	7593	7604	7615	7627	7638	7649	2	4	6	8	9
50	0.7660	7672	7683	7694	7705	7716	7727	7738	7749	7760	2	4	6	7	9
51	0.7771	7782	7793	7804	7815	7826	7837	7848	7859	7869	2	4	5	7	9
52	0.7880	7891	7902	7912	7923	7934	7944	7955	7965	7976	2	4	5	7	9
53	0.7986	7997	8007	8018	8028	8039	8049	8059	8070	8080	2	3	5	7	9
54	0.8090	8100	8111	8121	8131	8141	8151	8161	8171	8181	2	3	5	7	8
55	0.8192	8202	8211	8221	8231	8241	8251	8261	8271	8281	2	3	5	7	8
56	0.8290	8300	8310	8320	8329	8339	8348	8358	8368	8377	2	3	5	6	8
57	0.8387	8396	8406	8415	8425	8434	8443	8453	8462	8471	2	3	5	6	8
58	0.8480	8490	8499	8508	8517	8526	8536	8545	8554	8563	2	3	5	6	8
59	0.8572	8581	8590	8599	8607	8616	8625	8634	8643	8652	1	3	4	6	7
60	0.8660	8669	8678	8686	8695	8704	8712	8721	8729	8738	1	3	4	6	7
61	0.8746	8755	8763	8771	8780	8788	8796	8805	8813	8821	1	3	4	6	7
62	0.8829	8838	8846	8854	8862	8870	8878	8886	8894	8902	1	3	4	5	7
63	0.8910	8918	8926	8934	8942	8949	8957	8965	8973	8980	1	3	4	5	6
64	0.8988	8996	9003	9011	9018	9026	9033	9041	9048	9056	1	2	4	5	6
65	0.9063	9070	9078	9085	9092	9100	9107	9114	9121	9128	1	2	4	5	6
66	0.9135	9143	9150	9157	9164	9171	9178	9184	9191	9198	1	2	3	5	6
67	0.9205	9212	9219	9225	9232	9239	9245	9252	9259	9265	1	2	3	4	6
68	0.9272	9278	9285	9291	9298	9304	9311	9317	9323	9330	1	2	3	4	5
69	0.9336	9342	9348	9354	9361	9367	9373	9379	9385	9391	1	2	3	4	5
70	0.9397	9403	9409	9415	9421	9426	9432	9438	9444	9449	1	2	3	4	5
71	0.9455	9461	9466	9472	9478	9483	9489	9494	9500	9505	1	2	3	4	5
72	0.9511	9516	9521	9527	9532	9537	9542	9548	9553	9558	1	2	3	3	4
73	0.9563	9568	9573	9578	9583	9588	9593	9598	9603	9608	1	2	2	3	4
74	0.9613	9617	9622	9627	9632	9636	9641	9646	9650	9655	1	2	2	3	4
75	0.9659	9664	9668	9673	9677	9681	9686	9690	9694	9699	1	1	2	3	4
76	0.9703	9707	9711	9715	9720	9724	9728	9732	9736	9740	1	1	2	3	3
77	0.9744	9748	9751	9755	9759	9763	9767	9770	9774	9778	1	1	2	2	3
78	0.9781	9785	9789	9792	9796	9799	9803	9806	9810	9813	1	1	2	2	3
79	0.9816	9820	9823	9826	9829	9833	9836	9839	9842	9845	1	1	2	2	3
80	0.9848	9851	9854	9857	9860	9863	9866	9869	9871	9874	0	1	1	2	2
81	0.9877	9880	9882	9885	9888	9890	9893	9895	9898	9900	0	1	1	2	2
82	0.9903	9905	9907	9910	9912	9914	9917	9919	9921	9923	0	1	1	1	2
83	0.9925	9928	9930	9932	9934	9936	9938	9940	9942	9943	0	1	1	1	2
84	0.9945	9947	9949	9951	9952	9954	9956	9957	9959	9960	0	1	1	1	1
85	0.9962	9963	9965	9966	9968	9969	9971	9972	9973	9974	0	0	1	1	1
86	0.9976	9977	9978	9979	9980	9981	9982	9983	9984	9985	0	0	0	1	1
87	0.9986	9987	9988	9989	9990	9990	9991	9992	9993	9993	0	0	0	1	1
88	0.9994	9995	9995	9996	9996	9997	9997	9997	9998	9998	0	0	0	0	0
89	0.9998	9999	9999	9999	0.9999	1.0000	0000	0000	0000	0000	0	0	0	0	0
	0′	6′	12′	18′	24′	30′	36′	42′	48′	54′	1	2′	3	4	5

NATURAL COSINES

Proportional Parts
Subtract

	0′	6′	12′	18′	24′	30′	36′	42′	48′	54′	1′	2′	3′	4′	5′
0°	1·0000	0000	0000	0000	0000	1·0000	0·9999	9999	9999	9999	0	0	0	0	0
1	0·9998	9998	9998	9997	9997	9997	9996	9996	9995	9995	0	0	0	0	0
2	0·9994	9993	9993	9992	9991	9990	9990	9989	9988	9987	0	0	0	0	1
3	0·9986	9985	9984	9983	9982	9981	9980	9979	9978	9977	0	0	0	1	1
4	0·9976	9974	9973	9972	9971	9969	9968	9966	9965	9963	0	0	1	1	1
5	0·9962	9960	9959	9957	9956	9954	9952	9951	9949	9947	0	1	1	1	1
6	0·9945	9943	9942	9940	9938	9936	9934	9932	9930	9928	0	1	1	1	2
7	0·9925	9923	9921	9919	9917	9914	9912	9910	9907	9905	0	1	1	1	2
8	0·9903	9900	9898	9895	9893	9890	9888	9885	9882	9880	0	1	1	2	2
9	0·9877	9874	9871	9869	9866	9863	9860	9857	9854	9851	0	1	1	2	2
10	0·9848	9845	9842	9839	9836	9833	9829	9826	9823	9820	1	1	2	2	3
11	0·9816	9813	9810	9806	9803	9799	9796	9792	9789	9785	1	1	2	2	3
12	0·9781	9778	9774	9770	9767	9763	9759	9755	9751	9748	1	1	2	2	3
13	0·9744	9740	9736	9732	9728	9724	9720	9715	9711	9707	1	1	2	3	3
14	0·9703	9699	9694	9690	9686	9681	9677	9673	9668	9664	1	1	2	3	4
15	0·9659	9655	9650	9646	9641	9636	9632	9627	9622	9617	1	2	2	3	4
16	0·9613	9608	9603	9598	9593	9588	9583	9578	9573	9568	1	2	2	3	4
17	0·9563	9558	9553	9548	9542	9537	9532	9527	9521	9516	1	2	3	3	4
18	0·9511	9505	9500	9494	9489	9483	9478	9472	9466	9461	1	2	3	4	5
19	0·9455	9449	9444	9438	9432	9426	9421	9415	9409	9403	1	2	3	4	5
20	0·9397	9391	9385	9379	9373	9367	9361	9354	9348	9342	1	2	3	4	5
21	0·9336	9330	9323	9317	9311	9304	9298	9291	9285	9278	1	2	3	4	5
22	0·9272	9265	9259	9252	9245	9239	9232	9225	9219	9212	1	2	3	4	6
23	0·9205	9198	9191	9184	9178	9171	9164	9157	9150	9143	1	2	3	5	6
24	0·9135	9128	9121	9114	9107	9100	9092	9085	9078	9070	1	2	4	5	6
25	0·9063	9056	9048	9041	9033	9026	9018	9011	9003	8996	1	2	4	5	6
26	0·8988	8980	8973	8965	8957	8949	8942	8934	8926	8918	1	3	4	5	6
27	0·8910	8902	8894	8886	8878	8870	8862	8854	8846	8838	1	3	4	5	7
28	0·8829	8821	8813	8805	8796	8788	8780	8771	8763	8755	1	3	4	6	7
29	0·8746	8738	8729	8721	8712	8704	8695	8686	8678	8669	1	3	4	6	7
30	0·8660	8652	8643	8634	8625	8616	8607	8599	8590	8581	1	3	4	6	7
31	0·8572	8563	8554	8545	8536	8526	8517	8508	8499	8490	2	3	5	6	8
32	0·8480	8471	8462	8453	8443	8434	8425	8415	8406	8396	2	3	5	6	8
33	0·8387	8377	8368	8358	8348	8339	8329	8320	8310	8300	2	3	5	6	8
34	0·8290	8281	8271	8261	8251	8241	8231	8221	8211	8202	2	3	5	7	8
35	0·8192	8181	8171	8161	8151	8141	8131	8121	8111	8100	2	3	5	7	8
36	0·8090	8080	8070	8059	8049	8039	8028	8018	8007	7997	2	3	5	7	9
37	0·7986	7976	7965	7955	7944	7934	7923	7912	7902	7891	2	4	5	7	9
38	0·7880	7869	7859	7848	7837	7826	7815	7804	7793	7782	2	4	5	7	9
39	0·7771	7760	7749	7738	7727	7716	7705	7694	7683	7672	2	4	6	7	9
40	0·7660	7649	7638	7627	7615	7604	7593	7581	7570	7559	2	4	6	8	9
41	0·7547	7536	7524	7513	7501	7490	7478	7466	7455	7443	2	4	6	8	10
42	0·7431	7420	7408	7396	7385	7373	7361	7349	7337	7325	2	4	6	8	10
43	0·7314	7302	7290	7278	7266	7254	7242	7230	7218	7206	2	4	6	8	10
44	0·7193	7181	7169	7157	7145	7133	7120	7108	7096	7083	2	4	6	8	10
	0′	6′	12′	18′	24′	30′	36′	42′	48′	54′	1′	2′	3′	4′	5′

NATURAL COSINES

Proportional Parts Subtract

	0′	6′	12′	18′	24′	30′	36′	42′	48′	54′	1′	2′	3′	4′	5′
45°	0.7071	7059	7046	7034	7022	7009	6997	6984	6972	6959	2	4	6	8	10
46	0.6947	6934	6921	6909	6896	6884	6871	6858	6845	6833	2	4	6	8	11
47	0.6820	6807	6794	6782	6769	6756	6743	6730	6717	6704	2	4	6	9	11
48	0.6691	6678	6665	6652	6639	6626	6613	6600	6587	6574	2	4	6	9	11
49	0.6561	6547	6534	6521	6508	6494	6481	6468	6455	6441	2	4	7	9	11
50	0.6428	6414	6401	6388	6374	6361	6347	6334	6320	6307	2	4	7	9	11
51	0.6293	6280	6266	6252	6239	6225	6211	6198	6184	6170	2	5	7	9	11
52	0.6157	6143	6129	6115	6101	6088	6074	6060	6046	6032	2	5	7	9	12
53	0.6018	6004	5990	5976	5962	5948	5934	5920	5906	5892	2	5	7	9	12
54	0.5878	5864	5850	5835	5821	5807	5793	5779	5764	5750	2	5	7	9	12
55	0.5736	5721	5707	5693	5678	5664	5650	5635	5621	5606	2	5	7	10	12
56	0.5592	5577	5563	5548	5534	5519	5505	5490	5476	5461	2	5	7	10	12
57	0.5446	5432	5417	5402	5388	5373	5358	5344	5329	5314	2	5	7	10	12
58	0.5299	5284	5270	5255	5240	5225	5210	5195	5180	5165	2	5	7	10	12
59	0.5150	5135	5120	5105	5090	5075	5060	5045	5030	5015	2	5	8	10	12
60	0.5000	4985	4970	4955	4939	4924	4909	4894	4879	4863	3	5	8	10	13
61	0.4848	4833	4818	4802	4787	4772	4756	4741	4726	4710	3	5	8	10	13
62	0.4695	4679	4664	4648	4633	4617	4602	4586	4571	4555	3	5	8	10	13
63	0.4540	4524	4509	4493	4478	4462	4446	4431	4415	4399	3	5	8	10	13
64	0.4384	4368	4352	4337	4321	4305	4289	4274	4258	4242	3	5	8	11	13
65	0.4226	4210	4195	4179	4163	4147	4131	4115	4099	4083	3	5	8	11	13
66	0.4067	4051	4035	4019	4003	3987	3971	3955	3939	3923	3	5	8	11	13
67	0.3907	3891	3875	3859	3843	3827	3811	3795	3778	3762	3	5	8	11	13
68	0.3746	3730	3714	3697	3681	3665	3649	3633	3616	3600	3	5	8	11	14
69	0.3584	3567	3551	3535	3518	3502	3486	3469	3453	3437	3	5	8	11	14
70	0.3420	3404	3387	3371	3355	3338	3322	3305	3289	3272	3	5	8	11	14
71	0.3256	3239	3223	3206	3190	3173	3156	3140	3123	3107	3	6	8	11	14
72	0.3090	3074	3057	3040	3024	3007	2990	2974	2957	2940	3	6	8	11	14
73	0.2924	2907	2890	2874	2857	2840	2823	2807	2790	2773	3	6	8	11	14
74	0.2756	2740	2723	2706	2689	2672	2656	2639	2622	2605	3	6	8	11	14
75	0.2588	2571	2554	2538	2521	2504	2487	2470	2453	2436	3	6	8	11	14
76	0.2419	2402	2385	2368	2351	2334	2317	2300	2284	2267	3	6	8	11	14
77	0.2250	2232	2215	2198	2181	2164	2147	2130	2113	2096	3	6	9	11	14
78	0.2079	2062	2045	2028	2011	1994	1977	1959	1942	1925	3	6	9	11	14
79	0.1908	1891	1874	1857	1840	1822	1805	1788	1771	1754	3	6	9	11	14
80	0.1736	1719	1702	1685	1668	1650	1633	1616	1599	1582	3	6	9	11	14
81	0.1564	1547	1530	1513	1495	1478	1461	1444	1426	1409	3	6	9	11	14
82	0.1392	1374	1357	1340	1323	1305	1288	1271	1253	1236	3	6	9	12	14
83	0.1219	1201	1184	1167	1149	1132	1115	1097	1080	1063	3	6	9	12	14
84	0.1045	1028	1011	0993	0976	0958	0941	0924	0906	0889	3	6	9	12	14
85	0.0872	0854	0837	0819	0802	0785	0767	0750	0732	0715	3	6	9	12	14
86	0.0698	0680	0663	0645	0628	0610	0593	0576	0558	0541	3	6	9	12	15
87	0.0523	0506	0489	0471	0454	0436	0419	0401	0384	0366	3	6	9	12	15
88	0.0349	0332	0314	0297	0279	0262	0244	0227	0209	0192	3	6	9	12	15
89	0.0175	0157	0140	0122	0105	0087	0070	0052	0035	0017	3	6	9	12	15
	0′	6′	12′	18′	24′	30′	36′	42′	48′	54′	1′	2′	3′	4′	5′

NATURAL TANGENTS

	0′	6′	12′	18′	24′	30′	36′	42′	48′	54′	1′	2′	3′	4′	5′
											Proportional Parts				
0°	0·0000	·0017	·0035	·0052	·0070	·0087	·0105	·0122	·0140	·0157	3	6	9	12	15
1	0·0175	·0192	·0209	·0227	·0244	·0262	·0279	·0297	·0314	·0332	3	6	9	12	15
2	0·0349	·0367	·0384	·0402	·0419	·0437	·0454	·0472	·0489	·0507	3	6	9	12	15
3	0·0524	·0542	·0559	·0577	·0594	·0612	·0629	·0647	·0664	·0682	3	6	9	12	15
4	0·0699	·0717	·0734	·0752	·0769	·0787	·0805	·0822	·0840	·0857	3	6	9	12	15
5	0·0875	·0892	·0910	·0928	·0945	·0963	·0981	·0998	·1016	·1033	3	6	9	12	15
6	0·1051	·1069	·1086	·1104	·1122	·1139	·1157	·1175	·1192	·1210	3	6	9	12	15
7	0·1228	·1246	·1263	·1281	·1299	·1317	·1334	·1352	·1370	·1388	3	6	9	12	15
8	0·1405	·1423	·1441	·1459	·1477	·1495	·1512	·1530	·1548	·1566	3	6	9	12	15
9	0·1584	·1602	·1620	·1638	·1655	·1673	·1691	·1709	·1727	·1745	3	6	9	12	15
10	0·1763	·1781	·1799	·1817	·1835	·1853	·1871	·1890	·1908	·1926	3	6	9	12	15
11	0·1944	·1962	·1980	·1998	·2016	·2035	·2053	·2071	·2089	·2107	3	6	9	12	15
12	0·2126	·2144	·2162	·2180	·2199	·2217	·2235	·2254	·2272	·2290	3	6	9	12	15
13	0·2309	·2327	·2345	·2364	·2382	·2401	·2419	·2438	·2456	·2475	3	6	9	12	15
14	0·2493	·2512	·2530	·2549	·2568	·2586	·2605	·2623	·2642	·2661	3	6	9	12	16
15	0·2679	·2698	·2717	·2736	·2754	·2773	·2792	·2811	·2830	·2849	3	6	9	13	16
16	0·2867	·2886	·2905	·2924	·2943	·2962	·2981	·3000	·3019	·3038	3	6	9	13	16
17	0·3057	·3076	·3096	·3115	·3134	·3153	·3172	·3191	·3211	·3230	3	6	9	13	16
18	0·3249	·3269	·3288	·3307	·3327	·3346	·3365	·3385	·3404	·3424	3	6	10	13	16
19	0·3443	·3463	·3482	·3502	·3522	·3541	·3561	·3581	·3600	·3620	3	6	10	13	16
20	0·3640	·3659	·3679	·3699	·3719	·3739	·3759	·3779	·3799	·3819	3	6	10	13	17
21	0·3839	·3859	·3879	·3899	·3919	·3939	·3959	·3979	·4000	·4020	3	7	10	13	17
22	0·4040	·4061	·4081	·4101	·4122	·4142	·4163	·4183	·4204	·4224	3	7	10	14	17
23	0·4245	·4265	·4286	·4307	·4327	·4348	·4369	·4390	·4411	·4431	3	7	10	14	17
24	0·4452	·4473	·4494	·4515	·4536	·4557	·4578	·4599	·4621	·4642	4	7	11	14	18
25	0·4663	·4684	·4706	·4727	·4748	·4770	·4791	·4813	·4834	·4856	4	7	11	14	18
26	0·4877	·4899	·4921	·4942	·4964	·4986	·5008	·5029	·5051	·5073	4	7	11	15	18
27	0·5095	·5117	·5139	·5161	·5184	·5206	·5228	·5250	·5272	·5295	4	7	11	15	18
28	0·5317	·5339	·5362	·5384	·5407	·5430	·5452	·5475	·5498	·5520	4	8	11	15	19
29	0·5543	·5566	·5589	·5612	·5635	·5658	·5681	·5704	·5727	·5750	4	8	12	15	19
30	0·5774	·5797	·5820	·5844	·5867	·5891	·5914	·5938	·5961	·5985	4	8	12	16	20
31	0·6009	·6032	·6056	·6080	·6104	·6128	·6152	·6176	·6200	·6224	4	8	12	16	20
32	0·6249	·6273	·6297	·6322	·6346	·6371	·6395	·6420	·6445	·6469	4	8	12	16	20
33	0·6494	·6519	·6544	·6569	·6594	·6619	·6644	·6669	·6694	·6720	4	8	13	17	21
34	0·6745	·6771	·6796	·6822	·6847	·6873	·6899	·6924	·6950	·6976	4	9	13	17	21
35	0·7002	·7028	·7054	·7080	·7107	·7133	·7159	·7186	·7212	·7239	4	9	13	18	22
36	0·7265	·7292	·7319	·7346	·7373	·7400	·7427	·7454	·7481	·7508	5	9	14	18	23
37	0·7536	·7563	·7590	·7618	·7646	·7673	·7701	·7729	·7757	·7785	5	9	14	18	23
38	0·7813	·7841	·7869	·7898	·7926	·7954	·7983	·8012	·8040	·8069	5	10	14	19	24
39	0·8098	·8127	·8156	·8185	·8214	·8243	·8273	·8302	·8332	·8361	5	10	15	20	24
40	0·8391	·8421	·8451	·8481	·8511	·8541	·8571	·8601	·8632	·8662	5	10	15	20	25
41	0·8693	·8724	·8754	·8785	·8816	·8847	·8878	·8910	·8941	·8972	5	10	16	21	26
42	0·9004	·9036	·9067	·9099	·9131	·9163	·9195	·9228	·9260	·9293	5	11	16	21	26
43	0·9325	·9358	·9391	·9424	·9457	·9490	·9523	·9556	·9590	·9623	6	11	17	22	28
44	0·9657	·9691	·9725	·9759	·9793	·9827	·9861	·9896	·9930	·9965	6	11	17	23	29
	0′	6′	12′	18′	24′	30′	36′	42′	48′	54′	1′	2′	3′	4′	5′

NATURAL TANGENTS

	0′	6′	12′	18′	24′	30′	36′	42′	48′	54′	1′	2′	3′	4′	5′
											Proportional Parts				
45°	1·0000	·0035	·0070	·0105	·0141	·0176	·0212	·0247	·0283	·0319	6	12	18	24	30
46	1·0355	·0392	·0428	·0464	·0501	·0538	·0575	·0612	·0649	·0686	6	12	18	25	31
47	1·0724	·0761	·0799	·0837	·0875	·0913	·0951	·0990	·1028	·1067	6	13	19	25	32
48	1·1106	·1145	·1184	·1224	·1263	·1303	·1343	·1383	·1423	·1463	7	13	20	27	33
49	1·1504	·1544	·1585	·1626	·1667	·1708	·1750	·1792	·1833	·1875	7	14	21	28	34
50	1·1918	·1960	·2002	·2045	·2088	·2131	·2174	·2218	·2261	·2305	7	14	22	29	36
51	1·2349	·2393	·2437	·2482	·2527	·2572	·2617	·2662	·2708	·2753	8	15	23	30	38
52	1·2799	·2846	·2892	·2938	·2985	·3032	·3079	·3127	·3175	·3222	8	16	24	31	39
53	1·3270	·3319	·3367	·3416	·3465	·3514	·3564	·3613	·3663	·3713	8	16	25	33	41
54	1·3764	·3814	·3865	·3916	·3968	·4019	·4071	·4124	·4176	·4229	9	17	26	34	43
55	1·4281	·4335	·4388	·4442	·4496	·4550	·4605	·4659	·4715	·4770	9	18	27	36	45
56	1·4826	·4882	·4938	·4994	·5051	·5108	·5166	·5224	·5282	·5340	10	19	29	38	48
57	1·5399	·5458	·5517	·5577	·5637	·5697	·5757	·5818	·5880	·5941	10	20	30	40	50
58	1·6003	·6066	·6128	·6191	·6255	·6319	·6383	·6447	·6512	·6577	11	21	32	43	53
59	1·6643	·6709	·6775	·6842	·6909	·6977	·7045	·7113	·7182	·7251	11	23	34	45	57
60	1·7321	·7391	·7461	·7532	·7603	·7675	·7747	·7820	·7893	·7966	12	24	36	48	60
61	1·8040	·8115	·8190	·8265	·8341	·8418	·8495	·8572	·8650	·8728	13	26	38	51	64
62	1·8807	·8887	·8967	·9047	·9128	·9210	·9292	·9375	·9458	·9542	14	27	41	55	68
63	1·9626	·9711	·9797	·9883	1·9970	2·0057	·0145	·0233	·0323	·0413	15	29	44	58	73
64	2·0503	·0594	·0686	·0778	·0872	·0965	·1060	·1155	·1251	·1348	16	31	47	63	78
65	2·145	·154	·164	·174	·184	·194	·204	·215	·225	·236	2	3	5	7	8
66	2·246	·257	·267	·278	·289	·300	·311	·322	·333	·344	2	4	5	7	9
67	2·356	·367	·379	·391	·402	·414	·426	·438	·450	·463	2	4	6	8	10
68	2·475	·488	·500	·513	·526	·539	·552	·565	·578	·592	2	4	6	9	11
69	2·605	·619	·633	·646	·660	·675	·689	·703	·718	·733	2	5	7	9	12
70	2·747	·762	·778	·793	·808	·824	·840	·856	·872	·888	3	5	8	10	13
71	2·904	·921	·937	·954	·971	2·989	3·006	·024	·042	·060	3	6	9	12	14
72	3·078	·096	·115	·133	·152	·172	·191	·211	·230	·251	3	6	10	13	16
73	3·271	·291	·312	·333	·354	·376	·398	·420	·442	·465	4	7	11	14	18
74	3·487	·511	·534	·558	·582	·606	·630	·655	·681	·706	4	8	12	16	20
75	3·732	·758	·785	·812	·839	·867	·895	·923	·952	·981	5	9	14	19	23
76	4·011	·041	·071	·102	·134	·165	·198	·230	·264	·297	5	11	16	21	27
77	4·331	·366	·402	·437	·474	·511	·548	·586	·625	·665	6	12	19	25	31
78	4·705	·745	·787	·829	·872	·915	4·959	5·005	·050	·097	7	15	22	29	37
79	5·145	·193	·242	·292	·343	·396	·449	·503	·558	·614	9	18	26	35	44
80	5·671	·730	·789	·850	·912	5·976	6·041	·107	·174	·243	11	21	32	43	54
81	6·314	·386	·460	·535	·612	·691	·772	·855	6·940	7·026	13	27	40	54	67
82	7·115	·207	·300	·396	·495	·596	·700	·806	7·916	8·028	17	34	51	69	86
83	8·144	·264	·386	·513	·643	·777	8·915	9·058	·205	·357	23	46	68	91	114
84	9·514	9·677	9·845	10·019	10·199	10·385	10·579	10·780	10·988	11·205					
85	11·43	11·66	11·91	12·16	12·43	12·71	13·00	13·30	13·62	13·95	p.p cease to be sufficiently accurate				
86	14·30	14·67	15·06	15·46	15·89	16·35	16·83	17·34	17·89	18·46					
87	19·08	19·74	20·45	21·20	22·02	22·90	23·86	24·90	26·03	27·27					
88	28·64	30·14	31·82	33·69	35·80	38·19	40·92	44·07	47·74	52·08					
89	57·29	63·66	71·62	81·85	95·49	114·6	143·2	191·0	286·5	573·0					
	0′	6′	12′	18′	24′	30′	36′	42′	48′	54′	1′	2′	3′	4′	5′

TEACH YOURSELF BOOKS

CALCULUS

P. Abbott

This book has been written as a course in calculus both for those who have to study the subject on their own and for use in the classroom.

Although it is assumed that the reader has an understanding of the fundamentals of algebra, trigonometry and geometry, this course has been carefully designed for the beginner, taking him through a carefully graded series of lessons. Progressing from the elementary stages, the student should find that, on working through the course, he will have a sound knowledge of calculus which he can apply to other fields such as engineering.
A full length course in calculus, revised and updated, incorporating SI units throughout the text.

UNITED KINGDOM	**75p**
AUSTRALIA	**$2.45p***
NEW ZEALAND	**$2.45**
CANADA	**$2.95**

* recommended but not obligatory

ISBN 0 340 05536 7

TEACH YOURSELF BOOKS

PHYSICS

D. Bryant

This book offers a complete and unified guide to elementary physics for the interested layman who requires a modern approach to the subject.

In the form of a reader, *Physics* includes lucid discussions of molecular and atomic structure, forces, energy and waves, and of the behaviour of light, gases and electricity. The text is illustrated with numerous sketches and photographs, and no previous knowledge is assumed on the part of the reader beyond a familiarity with basic mathematics.

"An exceptionally readable account of physics up to 'O' level standard"

The Times Educational Supplement

UNITED KINGDOM	**75p**
AUSTRALIA	**$2.45***
NEW ZEALAND	**$2.45**
CANADA	**$2.95**

*** recommended but not obligatory**

ISBN 0 340 15251 6

HEALER'S QUEST

The world spun and dipped crazily about them. Zelia heard the booming steps, the angry growl, of the enraged populace. She saw the bobbing light of their torches, and then nothing. A blackness, a blankness, a void took them in, shielding them from all. And it was cold, so cold.

The next thing she knew they were rolling tip over tail in wet grass outside the walls of the city. Zelia performed another graceless somersault as Ares flopped next to her, dropping with a grunt. The healer raised herself gingerly from the ground and looked around. They had landed near their camp. Their horses whickered softly in the trees.

"I'm afraid to ask," Ares said in a deceptively matter-of-fact tone. "What sort of trick was that we just did?"

Pushing the hair out of her face with a sharp chatter of bells, Zelia stared at him a while before speaking. "I don't know. If we did what I think we did, taking a short cut between planes, it's impossible."

"That's nice." Ares stood to offer her a hand up, then stared at his open palm, blanched and turned away without comment.

Also in the Point Fantasy series:

Doom Sword
Peter Beere

Brog the Stoop
Joe Boyle

Dragonsbane
Patricia C. Wrede

Look out for:

Dragon's Bard
Susan Price

Wolf-Speaker
Tamora Pierce

Fire Wars
Jessica Palmer

Searching for Dragons
Patricia C. Wrede